The Story of the Pullman Car

GEORGE MORTIMER PULLMAN
1831—1897

The Story of the Pullman Car

BY

JOSEPH HUSBAND
Author of "America at Work" and "A Year in a Coal-Mine."

ILLUSTRATED

CHICAGO
A. C. McCLURG & CO.
1917

Published May, 1917

W. F. HALL PRINTING COMPANY, CHICAGO

ACKNOWLEDGMENT

OF THE many books from which information was drawn for the preparation of this volume the author wishes to make particular acknowledgment to *The Modern Railroad*, by Mr. Edward Hungerford, to the article "Railway Passenger Travel," by Mr. Horace Porter, published in *Scribner's Magazine*, September, 1888; and to *Contemporary American Biography*, as well as to the many newspapers and magazines from whose files information and extracts have been freely drawn.

J. H.

Chicago, April, 1917

CONTENTS

ILLUSTRATIONS

ILLUSTRATIONS

I

The Birth of Railroad Transportation

THE STORY OF THE PULLMAN CAR

CHAPTER I

THE BIRTH OF RAILROAD TRANSPORTATION

SINCE those distant days when man's migratory instinct first prompted him to find fresh hunting fields and seek new caves in other lands, human energy has been constantly employed in moving from place to place. The fear of starvation and other elementary causes prompted the earliest migrations. Conquest followed, and with increasing civilization came the establishment of constant intercourse between distant places for reasons that found existence in military necessity and commercial activity.

For centuries the sea offered the easiest highway, and the fleets of Greece and Rome carried the culture and commerce of the day to relatively great distances. Then followed the natural development of land communication, and at once arose the necessity not only for vehicles of transportation but for suitable roads over which they might pass with comfort, speed, and safety. Over the Roman roads the commerce of a great empire flowed in a tumultuous stream. Wheeled vehicles rumbled along the highways—heavy springless carts to carry the merchandise, lightly rolling carriages for the comfort of wealthy travelers.

The elementary principle still remains. The wheel and the paved way of Roman days correspond to the four-tracked route of level rails and the ponderous steel wheels of the mighty Mogul of today. In speed, scope, capacity, and comfort has the change been wrought.

The English stagecoach marked a sharp advance in the progress of passenger transportation. With frequent relays of fast horses a fair rate of speed was maintained, and comfort was to a degree effected by suspension springs of leather and by interior upholstery.

An interesting example of the height of luxury achieved by coach builders was the field carriage of the great Napoleon, which he used in the campaign of 1815. This carriage was captured by the English at Waterloo, and suffered the ignominious fate of being later exhibited in Madame Tussaud's waxwork show in London. The coach was a model of compactness, and contained a bedstead of solid steel so arranged that the occupant's feet rested in a box projecting beyond the front of the vehicle. Over the front windows was a roller blind, which, when pulled down admitted the air but excluded rain. The *secrétaire* was fitted up for Napoleon by Marie Louise, with nearly a hundred articles, including a magnificent breakfast service of gold, a writing desk, perfumes, and spirit lamp. In a recess at the bottom of the toilet box were two thousand gold napoleons, and on the top of the box were places for the imperial wardrobe, maps, telescopes, arms, liquor case, and a large silver chronometer by which the watches of the army were regulated. In such quarters did the great emperor jolt along over the execrable roads of Eastern Europe.

The stagecoach was established in England as a public conveyance early in the sixteenth century, and soon regular routes were developed throughout the country. Now for the first time a closed vehicle afforded travelers comparative comfort during their journey, and in the stagecoach with its definite schedule may be seen the early prototype of the modern passenger railroad. For three centuries the stagecoach slowly developed, and its popularity carried it to the continent and later to America. But by a radical invention transportation was suddenly transformed.

As early as the middle of the sixteenth century, and actually contemporaneous with the inception of the stagecoach, railways, or wagon-ways, had their origin. At first these primitive railways were built exclusively to serve the mining districts of England and consisted of wooden rails over which horse-drawn wagons might be moved with greater ease than over the rough and rutted roads.

The next step forward was brought about by the natural wear of the wheels on the wooden tracks, and consisted of a method of sheathing the rails with thin strips of iron. To avoid the buckling which soon proved a fault of this innovation, the first actual iron rails were cast in 1767 by the Colebrookdale Iron Works. These rails were about three feet in length and were flanged to keep the wagon wheels on the track.

For a number of years this simple type of railroad existed with little change. Over it freight alone was carried, and its natural limitations and high cost, compared with the transportation afforded by canals, seemed to hold but little promise for future expansion.

As early as 1804 Richard Trevithick had experimented with a steam locomotive, and in the ten years following other daring spirits endeavored to devise a practical application of the steam engine to the railway problem. But in 1814 George Stephenson's engine, the "Blucher," actually drew a train of eight loaded wagons, a total weight of thirty tons, at a speed of four miles an hour, and the age of the steam railroad had begun.

The first railroad to adopt steam as its motive power was the Stockton & Darlington, a "system" comprising three branches and a total of thirty-eight miles of track. On the advice of Stephenson, horse power was not adopted and several steam engines were built to afford the motive power. This road was opened on September 27, 1825, and preceded by a signalman on horseback a train of thirty-four vehicles weighing about ninety tons departed from the terminus with the applause of the amazed spectators.

The novelty of this new venture soon appealed so strongly to popular fancy that a month later a passenger coach was added, and a daily schedule between Stockton & Darlington was inaugurated.

This first railway carriage for the transportation of passengers was aptly named the "Experiment." Consisting of the body of a stagecoach it accommodated approximately twenty-five passengers, of which number six found accommodations within, while the others perched on the exterior and the roof of the vehicle. The fare for the trip was one shilling, and each passenger was permitted to carry fourteen pounds of baggage.

This early adaption of the stagecoach to the rapidly developed demand for passenger service necessitated the coinage of a new terminology, and it is not surprising that many words of stagecoach days remained. Among these "coach" is still preserved, and in England the engineer is still called the "driver"; the conductor, "guard"; locomotive attendants in the roundhouse, "hostlers," and the roundhouse tracks the "stalls."

In 1829 a prize of five hundred pounds ($2,500) for the best engine was offered by the directors of the Liverpool & Manchester Railway which was to be opened in the following year, and at the trial which was held in October three locomotives constructed on new and high-speed principles were entered. These were the "Rocket" by George and Robert Stephenson, the "Novelty" by John Braithwaite and John Erickson, and the "Sanspareil" by Timothy Hackworth. Due to the failure of the "Novelty" and the "Sanspareil" to complete the trial run and the successful performance of the "Rocket" in meeting the terms of the competition, the Stephensons were awarded the prize and received an order for seven additional locomotives. It is interesting to learn that on its initial trip the "Rocket" attained the unprecedented speed of twenty-five miles an hour.

In 1819 Benjamin Dearborn, of Boston, memorialized Congress in regard to "a mode of propelling wheel-carriages" for "conveying mail and passengers with such celerity as has never before been accomplished, and with complete security from robbery on the highway," by "carriages propelled by steam on level railroads, furnished with accommodations for passengers to take their meals and rest during the passage, as in packet; and that they be sufficiently high for persons to walk in without stooping." Congress, however, failed to call this

One of the earliest types of an American passenger car, drawn by Peter Cooper's experimental locomotive, "Tom Thumb." The tubular boilers of the locomotive were made from gun barrels.

memorial from the committee to which it was referred.

The development of the locomotive in America approximates its development in England. As early as 1827 four miles of track were laid between Quincy and Boston for the transportation of granite for the Bunker Hill Monument. Horses furnished the power, and the cars were drawn over wooden rails fastened to stone sleepers.

But reports of the wonders of the new English railways soon crossed the water, and in 1828 Horatio Allen was commissioned by the Delaware & Hudson Canal Company to purchase four locomotives in

"The Best Friend," the first locomotive built for actual service in America, hauling the first excursion train on the South Carolina Railroad, January 15, 1831.

England for use on its new line from Carbondale to Honesdale, Pennsylvania. Of these locomotives three were constructed by Foster, Rastrick, and Company, of Stourbridge, and one by George Stephenson. The first engine to arrive was the "Stourbridge Lion" and on the ninth of August,

1829, it was placed on the primitive wooden rails and, to the amazement of the spectators, Allen opened the throttle and in a cloud of smoke and hissing steam moved down the track at the prodigious speed of ten miles an hour.

One of the first railways in America was the old Mohawk & Hudson, which was chartered by an act of the New York legislature on April 17, 1826. The commissioners who were entrusted with the duty of organizing the company met for the purpose in the office of John Jacob Astor, in New York City, on July 29, 1826. One of their first official acts was to appoint Peter Heming chief engineer and send him to England to examine as to the feasibility of building a railroad. Mr. Heming's salary was fixed at $1,500 a year. In due course of time he returned from his European visit of observation and reported in favor of the project under consideration. Notwithstanding that he was absent six months, the expenses of his trip, charged by him to the company, were only $335.59. The road first used horse power and later on adopted steam for use in the day time, retaining horses, however, for night work. It was not deemed safe to use steam after dark. At first the trains consisted of one car each, in construction closely resembling the old-fashioned stagecoach.

The road connected the two towns of Albany and Schenectady, and was seventeen miles in length, but the portion operated by steam was only fourteen miles in length, horses being used on the

Early passenger cars, designed after the then prevalent type of horse coach. These cars were part of the train that ran on the formal opening of the Mohawk & Hudson Railroad (the first link of the New York Central System) on July 5, 1831.

inclined plane division from the top of one hill to the top of another.

Three years later a prize of $4,000 was offered by the Baltimore & Ohio Company for an American engine, and the following year a locomotive constructed by Davis and Gastner won the award by drawing fifteen tons at the rate of fifteen miles an hour. In 1832, Matthias W. Baldwin, founder of the Baldwin Locomotive Works in Philadelphia, designed his first locomotive, "Old Ironsides," for the Philadelphia, Germantown & Morristown Railroad; and soon after his second locomotive, the "E.

One of the first important improvements made by America in passenger cars was the introduction of the "bogie," or truck; the short curves of the American roads compelling the abandonment of the English type of four-wheeled car with rigid axles. The illustration shows a "bogie" car used on the Baltimore & Ohio Railroad in 1835.

L. Miller," was put in service on the South Carolina Railroad.

The first passenger service to be put in regular operation in America must be credited to the Charleston & Hamburg Railroad in the late fall of 1830.

The following year construction was begun on the Boston & Lowell Railroad, and in the same year a passenger train, previously mentioned, was put in service between Albany and Schenectady on the new Mohawk & Hudson Railroad.

The journal of Samuel Breck of Boston, affords an interesting glimpse of the conditions of contemporary railroad travel:

July 22, 1835. This morning at nine o'clock I took passage on a railroad car (from Boston) for Providence. Five or six other cars were attached to the locomotive, and uglier boxes I do not wish to travel in. They were made to stow away some thirty human beings, who sit cheek by jowl as best they can. Two poor fellows who were not much in the habit of making their toilet, squeezed me into a corner, while the hot sun drew from their garments a villainous compound of smells made up of salt fish, tar, and molasses. By and by just twelve—only twelve—bouncing factory girls were introduced, who were going on a party of pleasure to Newport. "Make room for the ladies!" bawled out the superintendent. "Come gentlemen, jump up on top; plenty of room there!" "I'm afraid of the bridge knocking my brains out," said a passenger. Some made one excuse, and some another. For my part, I flatly told him that since I had belonged to the corps of Silver Grays I had lost my gallantry and did not intend to move. The whole twelve were, however, introduced, and soon made themselves at home, sucking lemons, and eating green apples. . . . The rich and the poor, the educated and the ignorant, the polite and the vulgar, all herd together in this modern improvement in traveling . . . . and all this for the sake of doing very uncomfortably in two days what would be done delightfully in eight or ten.

To follow further the rapid development of the railroad in America would require many volumes.

Cars and locomotive in use on the Camden & Amboy Railroad in 1845. The cars were heated by wood stoves, the glass sash was stationary, and ventilation was possible only from a wooden-panelled window which could be raised a few inches.

As the canal building fever had seized the fancy of the American public in preceding years, so a similar enthusiasm was instantly kindled in the new railroad, and railroad travel became immediately the most popular diversion. In a relatively few years a web of track carried the smoking locomotive and its rumbling train of cars throughout the country. Crude, and lacking almost every convenience of the passenger coach of the present day, the early railway carriage served fully its new-born function. To the latter half of the century was reserved the development of those refinements which have rendered travel safe and comfortable, and the perfecting of those vast organizations that have placed in American hands the railroad supremacy of the world.

II

The Evolution of the Sleeping Car

CHAPTER II

THE EVOLUTION OF THE SLEEPING CAR

THE history of improved railway travel may be said to date from the year 1836, when the first sleeping car was offered to the traveling public. In the years which followed the actual inception of the railroad in the United States, railway travel was fraught with discomfort and inconvenience beyond the realization of the present day. Travel by canal boat had at least offered a relative degree of comfort, for here comfortable berths in airy cabins were provided as well as good meals and entertainment, but the locomotive, by its greatly increased speed over the plodding train of tow mules, instantly commanded the situation, and as the mileage of the pioneer roads increased, travel by boat proportionately languished.

The first passenger cars were little better than boxes mounted on wheels. Over the uneven track the locomotive dragged its string of little coaches, each smaller than the average street car of today.

From the engine a pall of suffocating smoke and glowing sparks swept back on the partially protected passengers. Herded like cattle they settled themselves as comfortably as possible on the stiff-backed, narrow benches. The cars were narrow and scant head clearance was afforded by the low, flat

Car in use in 1844 on the Michigan Central Railroad. Interesting as showing the rapid improvement in passenger coaches and how soon they approached the modern type of car in general appearance.

roof. From the dirt roadbed a cloud of dust blew in through open windows, in summer mingled with the wood smoke from the engine. In winter, a wood stove vitiated the air. Screens there were none. By night the dim light from flaring candles barely illuminated the cars.

In addition to these physical discomforts were added the dangers attending the operation of trains entirely unprotected by any of the safety devices now so essential to the modern railroad. No road boasted of a double track; there was no telegraph by which to operate the trains. The air brake was unknown until 1869, when George Westinghouse

Car constructed by M. P. and M. E. Green of Hoboken, New Jersey, in 1831 for the Camden & Amboy Railroad.

received his patent. The Hodge hand brake which was introduced in 1849 was but a poor improvement on the inefficient hand brake of the earlier days. The track was usually laid with earth ballast and the rail joints might be easily counted by the passengers as the cars pounded over them. Add to these discomforts the necessity of frequent changes from one short line to another when it was necessary for the passengers each time to purchase new tickets and personally pick out their baggage, due to the absence of coupon tickets and baggage checks, and the joys of the tourist may be realized.

As early as 1836 the officers of the Cumberland Valley Railroad of Pennsylvania installed a sleeping-car service between Harrisburg and Chambersburg. This first sleeping car was, as was later the first Pullman car, an adaption of an ordinary day coach to sleeping requirements. It was divided into four compartments in each of which three bunks were built against one side of the car, and in the rear of the car were provided a towel, basin, and water. No bed clothes were furnished and the weary passengers fully dressed reclined on rough mattresses with their overcoats or shawls drawn over them, doubtless marveling the while at the fruitfulness of modern invention. As time went on other similar cars, with berths arranged in three tiers on one side of the car, were adopted by various railroads, and occasional but in no manner fundamental improvements were made. Candles furnished the light, and the heat was supplied by box stoves burning wood or sometimes coal. For a number of years these makeshift cars found an appreciative patronage, and temporarily served the patrons of the road.

In the next ten years similar "bunk" cars were adopted by other railroads, but improvements were negligible and their only justification existed in the

Midnight in the old coaches previous to the introduction of the Pullman sleeping car. A night journey in those days was something to be dreaded.

ability of the passengers to recline at length during the long night hours. The innovation of bedding furnished by the railroad marked a slight progress, but the rough and none too clean sheets and blankets which the passengers were permitted to select from a closet in the end of the car, must have failed even in that day to give satisfaction to the fastidious.

But in the early fifties these very inconveniences fired the imagination of a young traveler who had bought a ticket on a night train between Buffalo and Westfield, and in his alert mind was inspired, as he tossed sleepless in his bunk, the first vision of a car that would revolutionize the railroad travel of the world and of a system that would present to the traveling public a mighty organization whose first purpose would be to contribute safety, convenience, luxury and a uniform and universal service from coast to coast.

George Mortimer Pullman was born in Brockton, Chautauqua County, New York, March 3, 1831. His early schooling was limited to the country schoolhouse, and at the age of fourteen his education was completed and he obtained employment at a salary of $40 a year in a small store in Westfield, New York, that supplied the neighboring farmers with their simple necessities. But the occupation of a country storekeeper failed to fix the restless mind of the boy, and three years later he packed his few possessions and moved to Albion, New York, where an older brother had developed a cabinet-making business.

CONVENIENCE OF THE NEW SLEEPING CARS.

(Timid Old Gent, who takes a berth in the Sleeping Car, listens.)

BRAKEMAN. "Jim, do you think the Millcreek Bridge safe to-night?"

CONDUCTOR. "If Joe cracks on the steam, I guess we'll get the Engine and Tender over all right. I'm going forward!"

Here Pullman found a wider field for his natural abilities, and at the same time acquired a knowledge of wood working and construction that was soon to afford the foundation for larger enterprises. During the ten years that followed there were times when the demands on the little shop of the Pullman brothers failed to afford sufficient occupation for the two young cabinet makers, and the younger brother, eager to improve his opportunities, began to accept outside contracts of various sorts. The state of New York had begun to widen the Erie Canal which passed through Albion. Clustered on its banks were numerous warehouses and other buildings, and the young man soon proved his ability to contract successfully for the necessary moving of these buildings back to the new banks of the canal. The venture was successful. An opportunity fortuitously created was seized, and not only was an increased livelihood secured, but the wider scope of this new activity gave the young man an increased confidence in himself on which to enlarge his future activities.

It was during these years that George M. Pullman experienced his first night travel and the hardships of the sleeping car accommodations. As Fulton and

Watt and Stephenson, in the crude steam engine of their time, saw the locomotive and marine engine of today, so in this bungling sleeper George M. Pullman saw the modern sleeping car and the vast system he was in time to originate. In his mind a score of ideas were immediately presented and on his return to Albion he discussed the possibility of their amplification with Assemblyman Ben Field, a warm friend in these early days.

The contracting business had increased Pullman's field of observation, it had stimulated his invention, it had accustomed him to the management of men. When the widening of the Erie Canal had been accomplished, the field for his new vocation was practically eliminated; and it was but natural that the ambition of youth could not be satisfied to return to the cabinet-making business. Westward lay the future. In the new town of Chicago, which had in so few years grown up at the foot of Lake Michigan, young men were already building world enterprises. Chicago, named from the wild onion that grew in the marsh lands about the winding river, offered promise of greatness. Its romantic growth seized the imagination of the youthful Albion contractor.

Naturally his first thought was to profit by his contracting experience, and again a happy chance favored him. Built on the low land behind the sand dunes and south of the sluggish river Chicago suffered from a lack of proper drainage. Mud choked the streets; cellars were wells of water after every rain. In 1855, the year of his arrival, Pullman made a contract to raise the level of certain of the city streets. It was a bold undertaking, but his confidence knew no hesitation, and the work was satisfactorily accomplished. Other contracts followed, and in a short time Pullman had built himself a substantial reputation and had raised a number of blocks of brick and stone buildings, including the famous Tremont House, to the new level.

Chicago in 1858 was a town of 100,000 population. Here Cyrus H. McCormick had built his reaper factory on the banks of the river. Here R. T. Crane was laying the small foundation for the mighty industry of future years. Here Marshall Field and Levi Z. Leiter were rising junior partners in their growing business, and here the future heads of the meat-packing industry were developing their mighty business. To the country boy from a New

York village, its muddy streets and rows of frame and brick buildings savored of a metropolis; in its naked newness he sensed the vital energy that was so soon to place it among the cities of the world.

But even during these years of untiring activity the thought of a radical improvement in railway car construction was constantly working in the brain of the young contractor, and in 1858 he determined to give his ideas the practical test. The story of this first application of these revolutionizing ideas to the railroad coaches then in use is best told in the words of Leonard Seibert, who was at that time an employee on the Chicago & Alton Railroad.

In 1858 Mr. Pullman came to Bloomington and engaged me to do the work of remodelling two Chicago & Alton coaches into the first Pullman sleeping-cars. The contract was that Mr. Pullman should make all necessary changes inside of the cars. After looking over the entire passenger car equipment of the road, which at that time constituted about a dozen cars, we selected Coaches Nos. 9 and 19. They were forty-four feet long, had flat roofs like box cars, single sash windows, of which there were fourteen on a side, the glass in each sash being only a little over one foot square. The roof was only a trifle over six feet from the floor of the car. Into this car we got ten sleeping-car sections, besides a linen locker and two washrooms — one at each end.

Early type of sleeping car. The traveler rarely removed more than his outer clothing, and oftentimes kept his boots on

The wood used in the interior finish was cherry. Mr. Pullman was anxious to get hickory, to stand the hard usage which it was supposed the cars would receive. I worked part of the summer of 1858, employing an assistant or two, and the cars went into service in the fall of 1858. There were no blue-prints or plans made for the remodelling of these first two sleeping-cars, and Mr. Pullman and I worked out the details and measurements as we came to them. The two cars cost Mr. Pullman not more than $2,000, or $1,000 each. They were upholstered in plush, lighted by oil lamps, heated with box stoves, and mounted on four-wheel trucks with iron wheels. There was no porter in those days; the brakeman made up the beds.

In the construction of these first sleeping cars Mr. Pullman introduced his invention of upper berth construction by means of which the upper berth might be closed in the day time and also serve as a receptacle for bedding. Other improvements and devices were worked out and tested, and from these first experiments were drawn the detailed plans from which the first cars entirely constructed by him were made. Although without technical training himself, Mr. Pullman was quick to recognize the necessity of skilled assistance to express and improve his embryonic ideas. To this end he soon established a small workshop, and employing a number of skilled mechanics set himself to the mastery of the problems which confronted him.

Another interesting personal reminiscence of the first days of the Pullman car is afforded by J. L. Barnes, who was in charge of the first car run from Bloomington to Chicago over the Chicago & Alton.

Mr. Pullman had an office on Madison Avenue just west of LaSalle Street and I boarded with a family very close to his office. I used to pass his office on my to meals, and having read in the paper that he was working on a sleeping car, one day I stopped in and made application to Mr. Pullman personally for a place as conductor. I gave him some references and called again and he said the references were all right and promised me the place. I made my first trip between Bloomington, Illinois, and Chicago on the night of September 1, 1859. I was twenty-two years old at the time. I wore no uniform and was attired in citizen's clothes. I wore a badge, that was all. One of my passengers was George M. Pullman, inventor of the sleeping car. . . . All the passengers were from Bloomington and there were no women on the car that night. The people of Bloomington, little reckoning that history was being made in their midst, did not come down to the station to see the Pullman car's first trip. There was no crowd, and the car, lighted by candles, moved away in solitary grandeur, if such it might be called. . . . . I remember on the first night I had to compel the passengers to take their boots off before they got into the berths. They wanted to keep them on—seemed afraid to take them off.

The first month business was very poor. People had been in the habit of sitting up all night in the straight back seats and they did not think much of trying to sleep while traveling. . . . . After I had made a few trips it was decided it did not pay to employ a Pullman conductor, and the car was placed in charge of the passenger conductor of the train which carried the sleeping car, and I was out of a job.

The first Pullman car was a primitive thing. Beside being lighted with candles it was heated by a stove at each end of the car. There were no carpets on the floor, and the interior of the car was arranged in this way: There were four upper and four lower berths. The backs of the seats were hinged and to make up the lower berth the porter merely dropped the back of the seat until it was level with the seat itself. Upon this he placed a mattress and blanket. There was no sheets. The upper berth was suspended from the ceiling of the car by ropes and pulleys attached to each of the four corners of the berth. The upper berths were constructed with iron rods running from the floor of the car to the roof, and during the day the berth was pulled up until it hugged the ceiling, there being a catch which held it up. At night it was suspended about half-way between the ceiling of the car and the floor. We used curtains in front and between all the berths. In the daytime one of the sections was used to store all the mattresses in. The car had a very low deck and was quite short. It had four wheel trucks and with the exception of the springs under it was similar to the freight car of today. The coupler was "link and pin;" we had no automatic brakes or couplers in those days. There was a very small toilet room in each end, only large enough for one person at a time. The wash basin was made of tin. The water for the wash basin came from the drinking can which had a faucet so that people could get a drink.

The two remodeled Chicago & Alton coaches were instantly accepted by the public, but despite their popularity, and the popularity of a third car which followed them, their originator considered them merely as experiments and in 1864 plans for the first actual Pullman car were completed which gave promise of a car radically different in its construction, appointments, and arrangement from anything heretofore attempted. Into this car Pullman resolutely cast the small capital that he had accumulated; in its success he placed the unswerving confidence that characterized his clear vision and indomitable determination to succeed. This model car was built in Chicago on the site of the present Union Station in a shed belonging to the Chicago & Alton Railroad, at a cost of $18,239.31, without its equipment, and almost a year was required before it was ready for service. Fully equipped and ready for service it represented an investment of $20,178.14. The "Pioneer" was the name chosen

J. L. Barnes, the first Pullman car conductor, whose reminiscences of that early period are quoted in this book

for its designation, and with the faith that other cars would soon be required the letter "A" was added, an indication that even Mr. Pullman's vision failed to anticipate the possible demand beyond the twenty-six letters of the alphabet.

Never before had such a car been seen; never had the wildest flights of fancy imagined such magnificence. Up to the building of the "Pioneer" $5,000 had represented the maximum that had ever been spent on a single railroad coach. It was unbelievable that this $18,000 investment could yield a remunerative return. The "Pioneer" had improved trucks with springs reinforced by blocks of solid rubber; it was a foot wider and two and a half feet higher than any car then in service, the additional height being necessary to accommodate the hinged upper berth of Mr. Pullman's invention. Combined with its unusual strength, weight, and solidity, its beauty and the artistic character of its furnishing and decoration were unprecedented. At one stride an advance of fifty years had been effected.

A further proof of Mr. Pullman's faith in the success of the "Pioneer" type of car is illustrated by the fact that due to its increased height and breadth the dimensions of station platforms and bridges at the time of its construction would not permit its passage over any existing railroad. It is said that these necessary changes were hastened in the spring of 1865 by the demand that the new "Pioneer" be attached to the funeral train which conveyed the body of President Lincoln from Chicago to Springfield. In this way one railroad was quickly adapted to the new requirements, and a few years later when the "Pioneer" was engaged to take General Grant on a trip from Detroit to his home town of Galena, Illinois, another route was opened to its passage.

Other roads soon made the necessary alterations to permit the passage of the "Pioneer" and its sister cars which were now under construction. The "Pioneer" had, by this time, won wide recognition and popularity, and a few months later was put in regular service on the Alton Road. So well were its dimensions calculated by Mr. Pullman that the "Pioneer" immediately became the model by which all railroad cars were measured, and to this day practically the only changes in dimensions have been in increased length.

To secure the continuous use of the "Pioneer" and other similar cars an agreement was effected between Mr. Pullman and the Chicago & Alton which marked the beginning of the vast system which today embraces the entire country and makes possible continuous and luxurious travel over a large number of distinct railroads. Thus in the space of a few years George M. Pullman not only evolved a type of railroad car luxurious and beautiful in design and embracing in its construction patents of great originality and ingenuity, but, in addition, evolved the rudimentary conception of a system by which passengers might be carried to any destination in cars of uniform construction, equipped for day or night travel, and served and protected by trained employees whose sole function is to provide for the passengers' safety, comfort, and convenience.

III

The Rise of a Great Industry

CHAPTER III

THE RISE OF A GREAT INDUSTRY

THE "Pioneer" had cost Mr. Pullman $20,000. Compared with the finest sleeping cars previously in use, it was clearly evident that a new development in luxurious travel had been accomplished. The best ordinary sleeping cars were considered expensive at $4,000. There was no more comparison between the "Pioneer" and its predecessors in comfort than in cost. But it remained to be seen what the public would think of it; whether they preferred luxury, comfort, and real service, to hardship, discomfort, and no service at a lower cost.

The new cars were larger, heavier, and more substantial than any previously constructed. Increased safety was one of their advantages. Moreover, they were far more beautiful from every aspect—artistically painted, richly decorated, and furnished with fittings for that day remarkable for their elaborate nature. They were universally admired, and quickly became the topic of interest among the traveling public. It is remarkable that at this early date the two features of the Pullman car which characterize it today—the features of safety and luxury—should have been so clearly defined.

It is human nature to accept each step forward as a new standard and it is characteristically American to refuse to accept an inferior article as soon as one superior is available, even if at greater cost. The "Pioneer" and its successors established such a standard, and immediately those accustomed and able to afford the increased rate required by the greater investment in the car, gladly and thankfully accepted it; while those whose nature usually inclines to haggling when the purse is touched, were convinced of the worth of the innovation by the assurance against disaster which the weight and strength of the Pullman cars assured.

The next car constructed by Mr. Pullman, after the "Pioneer" cost $24,000. And very soon after several additional cars were built at approximately the same cost, and were put in operation on the Michigan Central Railroad. Here was the great test. In these luxurious carriages and in the verdict of the traveling public rested the future of Mr.

Pullman's project. The question simply resolved itself to this: Did the public want them? In the old sleeping cars a berth had cost considerably less than it was necessary to charge for one in the new Pullman cars. In the mind of the inventor there was no question as to the verdict. The railroad authorities were equally certain the other way. They did not think the public would pay the extra sum.

There was but one way to decide, and Mr. Pullman made the suggestion that both Pullman cars and old style sleeping cars be operated on the same train at their respective prices. The results would show.

What happened is best described in the words of a contemporary writer.

> Mr. Pullman suggested that the matter be submitted to the decision of the traveling public. He proposed that the new cars, with their increased rate, be put on trains with the old cars at the cheaper rate. If the traveling public thought the beauty of finish, the increased comfort, and the safety of the new cars worth $2 per night, there were the $24,000 cars; if, on the other hand, they were satisfied with less attractive surroundings at a saving of 50 cents, the cheaper cars were at their disposal. It was a simple submission without argument of the plain facts

on both sides of the issue — in other words, an application of the good American doctrine of appealing to the people as the court of highest resort.

The decision came instantly and in terms which left no opening for discussion. The only travelers who rode in the old cars were those who were grumbling because they could not get berths in the new ones. After running practically empty for a few days, the cars in which the price for a berth was $1.50 were withdrawn from service, and Pullmans, wherein the two-dollar tariff prevailed, were substituted in their places, and this for the very potent reason, that the public insisted upon it. Nor did the results stop there. The Michigan Central Railway, charging an extra tariff of fifty cents per night as compared with other eastern lines, proved an aggressive competitor of those lines, not in spite of the extra charge, but because of it, and of the higher order of comfort and beauty it represented. Then followed a curious reversal of the usual results of competition. Instead of a levelling down to the cheaper basis on which all opposition was united, there was a levelling up to the standard on which the Pullman service was planted and on which it stood out single-handed and alone.

Within comparatively a short period all the Michigan Central's rival lines were forced by sheer pressure from the traveling public to withdraw the inferior and cheaper cars and meet the superior accommodations and the necessarily higher tariff. In other words, the inspiration of that key-note of vigorous ambition for excellence of the product itself, irrespective of immediate financial returns, which was struck with such emphasis in the building of the "Pioneer," and which ever since has rung through all

One of the first cars built by George M. Pullman

Interior of the car. (1) the car in the daytime showing wood stove and fuel box; (2) making up the berths. There were no end divisions, and a thin curtain only separated the berths

the Pullman work, was felt in the railroad world of the United States at that early date, just as it is even more commonly felt at the present time. At one bound it put the American railway passenger service in the leadership of all nations in that particular branch of progress, and has held it there ever since as an object lesson in the illustration of a broad and far-reaching principle.[1]

It will probably be interesting at this point to describe with some detail the Pullman car of this early period. In the *Daily Illinois State Register*, Springfield, May 26, 1865, appears an interesting description of one of the new Pioneer type of cars just installed on the Chicago & Alton Railroad.

To the train on the Chicago, Alton & St. Louis Railroad, which passed up at noon today, was attached one of Pullman's improved and beautiful sleeping carriages, containing a party of excursionists from the Garden City [Chicago], to whom the trip was complimentarily extended by the company of the road, and among whom was George M. Pullman, Esq., of Chicago, the patentee of the car. This carriage, which we had the pleasure of inspecting during the stay of the train at our depot, we found to be the most comfortable and complete in all its appurtenances, and decidedly superior in many respects to any similar carriage we have ever seen. It is fifty-four feet in length by ten in width, and was built at a cost of $18,000, the painting alone costing upwards of $500.

[1] *Contemporary American Biography*, p. 260.

Besides the berths, sufficient in number to accommodate upwards of a hundred passengers, there are four state rooms formed by folding doors, and so constructed with the berths that the whole can easily be thrown into one apartment. When the car is not used for sleeping purposes, as in the day, every appearance of a berth or a bed is concealed, and in their stead appear the most comfortable of seats.

Westlake's patent heating and ventilating apparatus is applied so that a constant current of pure and pleasant air is kept in circulation through the car. In fact, it was useless to attempt to enumerate, in so brief a notice, even a few of the many improvements which have been introduced by the patentees into the carriage, rendering it as they have, superior to any that we have ever inspected. To one fact, however, we will refer in this connection, as especially conducive to the comfort of the traveling public, viz., that a daily change of linen is made in the berths of this new carriage, thereby keeping them constantly clean and comfortable, and rendering the car much more attractive than are similar carriages where this is neglected. As we are informed by Mr. Pullman that these cars will hereafter be run on the St. Louis and Chicago line, we would especially direct the attention of travelers to the fact, and recommend them to investigate the matter of our notice for themselves.

Exactly how "upwards of a hundred passengers" could have been accommodated is hardly clear, but the enthusiasm of the reporter, fired perhaps by the luxury of clean linen for each berth each day, may account for this apparent exaggeration. In the *Illinois Journal*, another Springfield paper, of May 30, the reporter reduces the estimate of the capacity to fifty-two and comments with perhaps more detail on the decorative features of the car.

> We are reminded by a prophecy which we heard some three years since—that the time was not far distant when a radical change would be introduced in the manner of constructing railroad cars; the public would travel upon them with as much ease as though sitting in their parlors, and sleep and eat on board of them with more ease and comfort than it would be possible to do on a first-class steamer. We believed the words of the seer at the time, but did not think they were so near fulfillment until Friday last, when we were invited to the Chicago & Alton depot in this city to examine an improved sleeping-car, manufactured by Messrs. Field & Pullman, patentees, after a design by George M. Pullman, Esq., Chicago.

The writer describes his impressions of the interior. The absence of "mattresses or dingy curtains" by day, the beauty of the window curtains "looped in heavy folds," the "French plate mirrors suspended from the walls," as well as the "several beautiful chandeliers, with exquisitely ground shades" hanging from a ceiling "painted with chaste and elaborate design upon a delicately tinted azure ground," while the black walnut woodwork and "richest Brussels carpeting" make the picture complete. It is small wonder that the Pullman car excited admiration, and that its first appearance in the Illinois towns was probably recorded by similar editorial appreciation.

But perhaps one of the most interesting insights into the condition which the new Pullman cars were so quick to remedy, is found in the *Chicago Tribune*, June 20, 1865. After a veritable eulogy on the elegance and comfort of the Pullman car, the writer draws the following enviable contrast.

> It leaves to others to ticket the actual transit, so many miles for so much money, and comes in with its cars as the Ticket Agent of Comfort, sells you coupons to rest and ease by the way. So you wish to go through to New York or Baltimore, yourself, Belinda, Biddy and the baby, baskets, bundles, etc? You think of changes of cars by night, and rushes for seats for your party by day, of seats foul with the scrapings of dirty boots, of floors flowing with saliva, of coarse faces and coarse conversation, of seats you cannot recline in, of the ordinary discomforts of a long journey by rail!

It is small wonder that the new Pullman cars found an appreciative welcome!

George M. Pullman explaining details of car construction

RISE OF A GREAT INDUSTRY

In 1866 five Pullman sleeping cars were put in operation on the Chicago, Burlington & Quincy Railroad, and late in May an excursion for several hundred invited guests was given from Chicago to Aurora, Illinois, and return. The new cars were named, "Atlantic," "Pacific," "Aurora," "City of Chicago," and "Omaha." Occasioned by the comforts which this new equipment disclosed a current newspaper remarked:

> Pullman is a benefactor to his kind. The dreaded journey to New York becomes a mere holiday excursion in his delightful coaches, and, by the way, he will soon have a through line from Chicago to New York, in which a man need never leave his place from one city to the other.

The year 1867 marks the incorporation of Pullman's Palace Car Company, for the purpose of the manufacture and operation of sleeping cars. At the time of incorporation George M. Pullman owned all of the sleeping cars on the Michigan Central Railroad, Great Western [Canada] Railroad, and the New York Central Railroad lines, a grand total of forty-eight cars. In the operation of these cars he was ably assisted by his brother, A. B. Pullman, who held the office of general superintendent.

In forming the Pullman Company, the founder aspired to establish an organized system by which the traveling public might be enabled to travel in luxurious cars of uniform construction, adapted to both night and day requirements, without change between distant points, and over various distinct lines of railroads. In addition, such a service would provide the heretofore unknown asset of responsible employees to whose care might be entrusted women, children, and invalids. It was a service that was sorely needed, and indication pointed to its prompt acceptance by the railroads and the public.

In the same year a remarkable achievement in railroad travel was accomplished. Due to the different gauge tracks in use by the several railroads connecting Chicago and New York, the continuous passage of a car from one city to the other was impossible. But in 1867 the standardization of the gauge was effected by the completion of a third rail on the Great Western [Canada] Railroad, and to mark this opening of through communication, an excursion was arranged from Chicago to New York on the "Western World," the newest Pullman "hotel" sleeping car.

At this point it is interesting to note that the first "hotel car," the "President," was put in service by the Pullman Company in 1867 on the Great Western Railroad of Canada. The hotel car was a combination car, in reality a sleeping car with a kitchen built in at one end. The meals were served at tables placed in the sections. To the Pullman Company, accordingly, must be accorded the credit of first supplying to the public the service of meals on board a train. The success of the "President" led to the immediate construction of the "Western World" and its sister car "Kalamazoo." These cars, however, must not be confused with the dining car which was later developed from the "hotel car" by the Pullman Company, and to which the "hotel cars" rapidly gave place.

The *Detroit Commercial Advertiser* of June 1, 1867, comments:

> But the crowning glory of Mr. Pullman's invention is evinced in his success in supplying the car with a cuisine department containing a range where every variety of meats, vegetables and pastry may be cooked on the car, according to the best style of culinary art.

The following bill of fare illustrates the variety of edibles provided on this celebrated excursion.

MENU

OYSTERS

Raw 50
Fried and Roast..................... 60

COLD

Beef Tongue, Sugar-cured Ham,
Pressed Corned Beef, Sardines..... 40
Chicken Salad, Lobster Salad......... 50

BROILED

Beefsteak, with Potatoes............. 60
Mutton Chops, with Potatoes......... 60
Ham, with Potatoes................. 50

EGGS

Boiled, Fried, Scrambled, Omelette
Plain 40
Omelette with Rum.................. 50

Chow-Chow, Pickles

Welsh Rarebit 50
French Coffee 25
Tea 25

The excursion party left Chicago on April 8, 1867, and comfortably established in the "Western

World" arrived in Detroit the following day. At Detroit the river was crossed on the "great iron ferry boat," the first company of passengers that ever passed from Chicago to Canada without change of cars. On the new third rail of the Great Western, a speed of forty miles was often maintained for considerable periods. "The cars were decorated with American and British flags, symbolizing the union which is destined to take place between the United States and Canada. A train has just rolled by, the engine and passenger cars on the broad gauge, and freight cars from the East on the narrow gauge." So goes the journal of one of the passengers.

Large crowds visited the train at Rochester, Syracuse, and Utica, and at Albany, Erastus Corning telegraphed Commodore Vanderbilt that the car must be taken to New York, if possible, and the gauge of the Harlem road be taken for that purpose. The party arrived in New York on April 14. One of the purposes of sending the "Western World" to New York was that it might transport on its return trip, Dr. J. C. Durant, vice president of the Union Pacific Road, and a committee of directors, to examine a portion of their new transcontinental line which the contractors were ready to turn over. A member of the party describes the call on Dr. Durant in his office on Nassau Street and refers to the office as "probably the finest in New York, beautiful with paintings and statuary, and enlivened with the singing of birds."

Following the "Western World," the "hotel cars" were promptly put in service and regular through service was established between Chicago and eastern points. The new "City of Boston" and "City of New York" surpassed even the "Western World" in magnificence and were popularly reported to have exceeded $30,000 each in cost. These cars were known as "hotel cars" for the reason that each contained all the requirements for a protracted journey. The main body of the car was occupied by the berths and seats and at one end a kitchen and pantry provided the culinary service. The dining car, devoted entirely to restaurant purposes, was a second step which soon followed. The first dining car personally designed by Mr. Pullman was named the "Delmonico," and was operated on the Chicago & Alton in 1868.

One of the first Pullman cars in which meals were served

But it was in 1869 that the Pullman car made perhaps its greatest advance in the interest and confidence of the public for in that year the Union Pacific, building westward from the Missouri River at Omaha, met the Central Pacific, which built from San Francisco eastward. By their union a line was established between the two coasts of the continent, a slender thread of track which stretched for 1,848 miles through a practically uninhabited country. Almost simultaneously with the completion of the road there was put upon the rails one of the most superb trains ever turned out of the Pullman shops. Its journey to California and its reception there were in the nature of a progressive ovation. From that time forth the great population of the Pacific coast knew no train for long distance travel save a Pullman train, and would hear of no other. When people from California reached Chicago on their way eastward, the road over which Pullman cars ran got their patronage, and roads over which other cars were operated did not. Newspapers and magazines were awakened to studies of the Pullman cars and the Pullman system, and scores of printed pages were filled with the marvels of a journey to the

Pacific Ocean which was nothing more than a six days' sojourn in a luxurious hotel, past the windows of which there constantly flowed a great panorama of the American continent, thousands of miles in length and as wide as the eye could reach. Illustrated magazine articles which appeared telling the story of a trip to California had as many pictures of Pullman interiors as they had of the big trees or the Yosemite Valley. The effect of all this was far reaching. The great Pennsylvania line abandoned its own service and adopted the Pullman, and many other lines made application for inclusion in the Pullman system.

In May, 1870, the first through train from the Atlantic to the Pacific crossed the continent, engaged for a special excursion by the Boston Board of Trade, many distinguished Bostonians being numbered among the passengers. During the trip a daily newspaper entitled the *Trans-Continental* was published. In the issue of May 31, published on the sixth day out, as the train was crossing the summit of the Sierra Nevadas, an account is given of a meeting of the passengers in the smoking car, and resolutions passed by them were printed. The Hon. Alex

H. Rice presided at the meeting, and the resolutions were offered by Frank H. Peabody, a Boston banker, and seconded by Robert B. Forbes, another Bostonian.

Resolved, That we, the passengers of the Boston Board of Trade Pullman excursion train, the first through train from the Atlantic to the Pacific, having now been a week *en route* for San Francisco, and having had, during this period, ample opportunity to test the character and quality of the accommodations supplied for our journey, hereby express our entire satisfaction with the arrangements made by Mr. George M. Pullman, and our admiration of the skill and energy which have resulted in the construction, equipment and general management of this beautiful and commodious moving hotel.

Resolved, That we return our cordial thanks to Mr. Pullman for the very great pains taken by him beforehand to make the present journey safe and pleasurable; that we recognize the complete success which has followed all his efforts, and that we extend to him our sincere wishes for such a degree of prosperity to attend all his operations as will be proportionate to his merits as one of the most public-spirited, sagacious, and liberal railroad men of the present day.

Resolved, That we take pleasure in witnessing, as we journey from point to point, through all the Western States, the many evidences of Mr. Pullman's enterprise and the extent of his operations in the cars which we meet belonging to the Pullman Company, attached to the regular trains for the use of the public, or appropriated especially to private excursion parties, and we earnestly hope that there will be no delay in placing the elegant and homelike carriages upon the principal routes in the New England States, and we will do all in our power to accomplish this end.

The list of passengers on this notable excursion included:

Hon. Alex. H. Rice
Maj. Geo. P. Denny
Hon. J. M. S. Williams
James W. Bliss
Edward W. Kingsley
Frederick Allen and wife
H. S. Berry
Miss Josie W. Bliss
Hon. John B. Brown and wife
E. W. Burr and son
John L. Bremer
Geo. D. Baldwin and wife
Miss L. E. Billings
Chas. W. Brooks
M. S. Bolles
Alvah Crocker and wife
Mrs. F. Cunningham
Thomas Dana, Mrs. Thomas Dana, 2nd, Miss M. E. Dana
Mrs. Geo. P. Denny
Arthur B. Denny
Cyrus Dupee and wife
John H. Eastburn and wife
Robert B. Forbes and wife
Joshua Reed
J. S. Fogg
Mrs. E. E. Poole
Misses Farnsworth
Robert O. Fuller
J. Warren Faxon
N. W. Farwell and wife
Miss Mary E. Farwell
Miss Evelyn A. Farwell
Curtis Guild and wife
C. L. Harding and wife
Miss N. Harding
Edgar Harding
J. F. Hunnewell
J. F. Heustis
W. S. Houghton and wife
D. C. Holder and wife
Miss C. Harrington
A. L. Haskell and wife
Miss Alice J. Haley

J. M. Haskell and wife
H. O. Houghton and wife
John Humphrey
Hamilton A. Hill and wife
Benjamin James
C. F. Kittredge
Mrs. C. A. Kinglsey
Miss Addie P. Kinglsey
Miss Mary L. Kinglsey
Chas. S. Kendall
Miss M. C. Lovejoy
John Lewis
Jas. Longley and wife
Geo. Myrick and wife
Col. L. B. Marsh and wife
C. F. McClure and wife
Joseph McIntyre
Sterne Morse
Fulton Paul
F. H. Peabody, wife and servant
Miss F. Peabody
Miss L. Peabody
Master F. E. Peabody
Rev. E. G. Porter
Miss M. F. Prentiss
James W. Roberts and wife
Wm. Roberts
S. B. Rindge and wife
Master F. H. Rindge
J. M. B. Reynolds and wife
John H. Rice
Hon. Stephen Salisbury
M. S. Stetson and wife
D. R. Sortwell and wife
Alvin Sortwell
F. H. Shapleigh
T. Albert Taylor and wife
E. B. Towne
Lawson Valentine and wife
Miss Valentine
Rev. R. C. Waterston and wife
A. Williams
Dr. H. W. Williams and wife
N. D. Whitney and wife
Judge G. W. Warren
Geo. A. Wadley and wife
Henry T. Woods
Mrs. J. M. S. Williams
Miss E. M. Williams
Miss C. T. Williams
J. Bert Williams

In the next few years the Pullman Palace Car Company established manufacturing shops in

Detroit, and in 1875 a new "reclining-chair car," the first parlor car to be operated in the United States, was presented by Mr. Pullman to the public. For several years parlor cars of Pullman design and construction had been in satisfactory use on the Midland Railway, between London and Liverpool, England. The success of these cars promptly resulted in the construction of the "Maritana" for use in the United States. The chairs in this new car were heavily and richly upholstered and revolved on a swivel, on the same principle as the chairs in the parlor car of the present day.

The first parlor car, 1875

IV

The Pullman Car in Europe

CHAPTER IV

THE PULLMAN CAR IN EUROPE

A MODEST paragraph in many American newspapers in February, 1873, announced the momentous news that England was soon to enjoy the novelty of Pullman transportation — "The Midland Railway Company has entered into a contract with the Pullman Palace Car Company for the equipment of their road with American drawing room and sleeping coaches." The Midland was the longest and most important of three great railroads which started from London and extended to Liverpool and Scotland, transversing the rich central counties of England where so few years before the coach horn had sounded through the hills. The adoption of Pullman equipment by this prominent railroad was singularly conspicuous.

On February 15, 1873, at a "half-yearly meeting of the shareholders of the Midland Railway," Mr. Pullman personally addressed the officers of the company. It appears that Mr. Allport, the general manager of the Midland Railway, on a recent visit to the United States and Canada, had been greatly impressed by the accommodations afforded the traveling public, and had made a particular study of the Pullman cars. Acting on his advice the directors invited Mr. Pullman to England to appear before the meeting. Mr. Pullman proposed that the Midland Company should authorize the speedy construction of carriages particularly adapted to their requirements, and a motion was carried to authorize the construction of such cars on the basic Pullman principles. It was accordingly agreed that eighteen new cars should be constructed in America and shipped to England in August and that Mr. Pullman should return to England at that time to superintend their installation.

By the contract the Pullman Company agreed to furnish as many dining-room, drawing-room, and sleeping cars as the demands of the traveling public required, without charge to the road, its compensation being in the extra fare paid for use of the cars. The road, on the other hand, received its compensation in the free use of the cars, in return for which it guaranteed to the Pullman Company the exclusive right to furnish such cars for fifteen years. As in America, the porters, conductors, cooks, waiters and other attendants were hired by the Pullman Company. Two night trains and two day trains of American cars only, were to be put on at the start. The contract was not exclusive, and other English railroads watched with interest the working out of the American innovation.

The popularity of the Pullman car at home and abroad quite naturally inspired a host of imitators. Among the first was Colonel W. D. Mann, the proprietor of the *Mobile Register*, who designed a sleeping car embodying certain characteristic Pullman features, but divided transversely into compartments or "boudoirs," each entered directly from the sides, and connected by a private door permitting the passage of the attendant to and through the several compartments. Each compartment contained seats for four persons, which by night could be made up into beds. The design was ingenious but failed in many vital respects to compete with the greater comfort and roominess of the Pullman car.

As the Pullman car was the first sleeping car to be installed for regular service in England, so credit should be given to Colonel Mann for affording the first sleeping car for public service ever operated on the Continent. Mann's "Boudoir Cars" were installed on the Vienna and Munich line in 1873, and their favorable reception and popularity unquestionably went far to better the trying conditions of European travel.

Designed in America and introduced on the continent, the Mann boudoir cars enjoyed an almost unoccupied field in Europe, with the exception of England, where the railway managers had adopted the Pullman cars as their standard. The Mann car was developed to suit European railroads and European wants. A Belgian company was organized to introduce sleeping cars by contracts with railroad companies, somewhat like those of the Pullman Company in America. The Mann cars which were put in service in the United States between Boston and New York in 1883 were divided into eight compartments, some accommodating two persons, some four. The seats were arranged transversely instead of longitudinally. Due to their smaller passenger capacity a higher rate was necessarily charged than for Pullman accommodations.

Interior of a Pullman car used about 1880. Here a tendency to ornamentation begins to show. Note the low-backed seats

But exclusive possession of the Continental field was not left to Colonel Mann undisputed, for during the year 1875 Mr. Pullman established a shop at Turin, Italy, and under the direction of a Mr. A. Rapp, who was sent on from the Detroit works, a number of cars were constructed for use on through trains on the principal Italian lines. The following testimonial presented to Mr. Rapp at the conclusion of the work by the men who had been employed expresses, although in none too polished English, their appreciation of the work that had been provided them.

TO
PULLMAN ESQUIRE, THE GREAT INVENTOR
OF THE
SALOON COMFORTABLE CARRIAGES
AND
MASTER RAPP THE CIVIL ENGINEER, DIRECTOR
OF THE MANUFACTURE OF THE SAME
THE
ITALIAN WORKMEN
BEG TO UMILIATE.

Welcome, Welcome Master Pullman
The great inventor of the Saloon Carriages,
Italy will be thankful to the man
For now and ever, for ages and ages.

To Master Rapp we men are thankful.
Cause of his kindness and adviser sages,
Our hearts of true gladness is full:
And we shall remember him for ages.

Should Master Pullman ever succeed
To continue is work in Italy
What we wish to him indeed,
We hope to be chosen
To finish the work and work as a man,
To show our gratitude to Master Pullman.

FINO AND HIS FRIENDS.

Turin, 10 January 1876.

The appearance of the new Pullman cars in England created immediate and favorable comment, for not only were the cars radical in the service which they afforded, but their construction, following the advanced principles of American car building, offered sharp contrast to the less modern cars of English construction. From the most gorgeous first-class carriage down to the dumpiest begrimed coal car, all British railway conveyances rested on four iron wheels, placed in the position where Artemus Ward located the legs of the horse—one at each corner. Until the Pullman sleepers were introduced into Britain, the sight of a car resting on eight wheels was unprecedented, as no one thought of doubting the entire security from danger of a carriage with only four points of support. Indeed, the conservative Briton saw no more real necessity for a railway carriage having eight wheels than for a horse to have more than four legs.

Under arrangements with the Great Northern Railway, Pullman "dining room" carriages were put in service on November 1, 1879, between Leeds and King's Cross Station, London. Luncheon and dinner were served and the menu included "soups, fish, entrees, roast joints, puddings and fruits for dessert," a truly English bill of fare. The reception of this innovation is described by the *London Telegraph*, which concluded a comment on the dining car with this friendly suggestion:

If the British public can be brought to give this new refreshment-car system, just inaugurated by the Great Northern Railway, a fair trial, there will be another traveling infliction, besides Dyspepsia and Discontent, which will be speedily laid in the Red Sea. I mean the ghost of Ennui. Luncheon or dinner on board a Pullman palace-car will surely banish Boredom from railway journeys.

By the year 1879 Pullman sleeping and drawing room cars were in operation on three English and three Scotch lines, and at the invitation of the Italian Government, cordially responded to by the Pullman Palace Car Company, sleeping cars, similar to those in use in England on the Midland and Great Northern railways were put in weekly service between Brindisi and Bologna, in connection with the steamers of the Peninsula and Oriental Company. At Bologna the service was taken up by the Belgian "Societe Anonyme des Wagons Lits"—an interesting recognition by a foreign government of the superiority of the American railway carriages.

In 1888 "The Pullman Limited Express" began regular service on the London, Brighton, & South Coast Line, between Victoria Station and Brighton. Single cars of the American pattern had been running on this line for five or six years, but in this train for the first time the English public was offered a "solid Pullman" equipment. Four cars comprised the train—a parlor car, a drawing room car with ladies' boudoir and dining room, a restaurant car, and a smoking car, while a compartment at each end of the train next to the luggage compartment was provided for servants. On this train electric lighting was first employed by the Pullman Company for

illuminating railroad cars — a particular feature that received wide advertisement.

The London, Brighton, & South Coast Railway opened the New Year of 1889 with the first "vestibule" train that had ever greeted the eyes of foreign travelers. Three Pullman cars, "Princess," "Prince," and "Albert Victor," were regularly attached to a train of three first-class cars. The Pullman cars were built at the Pullman plant at Detroit, Michigan, and were shipped in sections to England. By this innovation Yankee genius again demonstrated its leadership, and the travelers of a distant nation profited by the genius and energy of an American inventor.

The Pullman Company, Limited, of England, existed as a property of the American company until the year 1906, when, due to the enormous development of the system in the United States, it was deemed wise for economic reasons to separate the two companies. But today the British company still proudly bears the name of Pullman, a tribute to the inventive genius, untiring energy, and wide vision of a country boy of the new world.

V

The Survival of the Fittest

CHAPTER V

THE SURVIVAL OF THE FITTEST

ONE of the most interesting elements in the history of the Pullman car and the Pullman Company is the story of imitation and competition which for a period after the foundation of the parent company thrived and later disappeared. The success of the Pullman car necessarily brought competition. It was wholesome that such competition should arise. If a car more convenient than the car of Mr. Pullman's invention could be devised, it was right that it should be given the test of public opinion. That no car constructed along different basic lines survived, established the right of the Pullman car to its preeminence. That certain cars patterned after Mr. Pullman's basic ideas, and in most cases directly infringing on his patents, received a degree of popularity again reflects creditably to the Pullman car.

Distinct from the innovations afforded by Pullman car construction, the universal service of the Company afforded the public a new service of equal value. Where formerly it was necessary for the traveler to change from car to car whenever and wherever one railroad connected with another line, the uniform service of the Pullman Company created a new and infinitely more desirable situation, for it was now possible to travel without inconvenience or interruption between practically any two points in the country regardless of the number of different railroads over whose tracks the traveler's ticket required passage. By competition, the value of such a service was tested; tested alike by the individual railroads and their patrons. That each and every competing company ultimately retired from the field, and that practically every railroad in the United States has today contracted with the Pullman Company for its standardized service, is tacit recognition to the worth of the service rendered.

There are still other reasons why the control of sleeping and parlor service should be delegated to a single company. Due to the vast area embraced by the boundaries of the United States and the wide range of climate which these boundaries contain, there are many railroads which require during certain months of the year a larger number of cars to trans-

More ornate interiors. (1) early Pullman parlor car; (2) old type Pullman sleeping car

port their through passengers than in others. Other roads require an equally great number of sleeping and parlor cars during other months, as for instance those roads which carry the winter tourists to the South and Southwest in winter as opposed to the roads which feel the peak of passenger travel in summer when the vacationists are headed for the Atlantic coast resorts or the northwestern mountains. Again, there are special occasions, like great conventions, when the railroads touching the convention city must have hundreds of sleeping cars above their normal needs.

Few railroads could afford to tie up capital in the cars required for such brief periods of demand; it would be an economic fallacy to pass the expense of the maintenance and constant replacement of such an equipment on to the public. To meet this situation is the mission of the Pullman Company.

Of the numerous sleeping car companies the Gates Sleeping Car Company was perhaps the earliest. This car was named after Mr. G. B. Gates, General Manager of the Lake Shore Road, and with the consolidation of the Hudson River Railroad and the New York Central in 1869, these cars, previously only operated on the Lake Shore, were put in the New York, Buffalo, Chicago service.

Among the various competitors of the Pullman Company, the Wagner Palace Car Company, which succeeded, in 1865, the New York Central Sleeping Car Company, and absorbed in 1869 the Gates Sleeping Car Company, developed by far the widest and most formidable competition and continued its service over the longest period. The underlying reasons for the strength of this competition lay primarily in the fact that the Wagner cars followed more closely the Pullman characteristics, and in fact the infringement of certain basic Pullman patents by the Wagner Company was a cause of frequent litigation over a period of many years. Webster Wagner, the founder of the Wagner Palace Car Company, began his career as a wagon maker. The first cars which he constructed had a single tier of berths, and the bedding was packed away by day in a closet at the end of the car. Commodore Vanderbilt backed Wagner and became interested in his company, a connection which gave Wagner invaluable assistance and a hold on the sleeping-car business of the lines controlled by the

The latest Pullman parlor car, showing simplicity of modern car decoration, combining quiet elegance with good taste and comfort

Vanderbilt interests, a connection which enabled him for many years to be a keen competitor of the Pullman Company.

Early in June, 1881, suit was brought by the Pullman Palace Car Company against the New York Central Sleeping Car Company and Webster Wagner, claiming $1,000,000 damages for infringement and use of patents in the construction and use of Wagner sleeping coaches. The bill stated that in 1870 the Wagner Company began building sleeping cars, and for several years its coaches ran only on the New York Central Railroad and its various branches. The company finding it impossible to build satisfactory cars without using the Pullman patents, contracted with the Pullman Company to use certain of its patented improvements. This arrangement was made with the distinct understanding that the Wagner Company was to run its cars only over the New York Central Railroad. For five years this arrangement was satisfactorily carried out. But in 1875 the Pullman Company's contract with the Michigan Central Railroad expired and the Wagner Company secured the contract to run the cars between Detroit and Chicago, thus making a through connection for the Vanderbilt lines between New York and Chicago.

By this new routing of the Wagner cars direct from New York to Chicago and the elimination of the Pullman cars from the Chicago and Detroit service, an opportunity offered for some other road to avail itself of the Pullman service and effect a through Pullman service between New York and Chicago.

The Erie was the road that grasped the opportunity. By arrangements with the Baltimore & Ohio and several other roads, through Erie trains between New York and Chicago, comprising Pullman hotel coaches, sleeping cars and drawing room cars were put in service on November 1, 1875. A circular published in Chicago announcing the new arrangement said:

> From the first of November, the Pullman hotel and drawing room coaches, for many years so popular on the Michigan Central line, will be withdrawn from that route, and with new and increased improvements will thereafter run exclusively on the Erie and Chicago line, forming the first and only Pullman hotel coach line between Chicago and New York.

The success of the new Erie Pullman coaches was immediately assured. The hotel cars especially were a great attraction. These were divided into two compartments, in one of which the kitchen was located, the other compartment being utilized as a sleeping car. First-class meals, including all manner of game and seasonable delicacies, were served on movable tables placed in the sections. In fact, the *New York Tribune*, in commenting on the new Pullman equipment, asked: "Should the Erie have a monopoly of such comforts? Why does not Wagner imitate or improve upon Pullman?"

These cars were nicknamed "French Flats."

All the modern conveniences of a first-class house are condensed into one of these hotels on wheels. The beds at night are put away to make room for spacious seats by day, between which a table is placed, covered with damask cloths and napkins folded in quaint devices, at which four may sit with ease. The whole car—a Pullman—is luxuriously fitted up, and one end is partitioned into a storeroom and kitchen; there is a smoking-room for lovers of the weed, and a separate toilet room for ladies. As the porter of the car blackens the boots, and there is a telegraph office at each stopping place, the waggish question of "Where is the barber shop?" is often made. But this may come, too, as last summer an excursion party of ladies and gentlemen took a hair-dresser with them over the Erie to Niagara Falls, and two or three ladies actually ***had their hair crimped*** while traveling thirty or forty miles an hour! At this time, while game is plenty in the West, the Pullmans, with their facilities, and two fast trains each way per day, are able to make a bill of fare and serve it in a style which would cause Delmonico to wring his hands in anguish. The service is on the European plan; that is, you pay for what you order, and we give the prices of the principal articles, to show at what a reasonable rate one can take a superior meal of fifty or a hundred miles long: Prairie chicken, pheasant, and woodcock, whole, $1; snipe, quail, golden plover and blue-winged teal, each 75 cents; venison, 60 cents; chicken, whole, 75 cents; cold tongue, ham, and corned beef, 30 cents; sardines, lobster, and broiled ham or bacon, 40 cents; mutton and lamb chops, veal cutlets, or half a chicken, 50 cents; sirloin steak, 50 cents, &c. Every traveler who has missed his dinner to catch a train will rejoice in knowing that a warm meal awaits him at the cars, and that he can wake up in the morning and choose his time for breakfast, instead of bolting it down at the twenty minutes' convenience of the railroad company.[1]

Some time prior to 1861 sleeping cars were being operated over the Camden & Amboy and Baltimore & Ohio railroads. These cars were known as "Knight" cars, after their designer, E. C. Knight. The "Knights" were built at a cost of about $7,000, and were regarded as the handsomest things on wheels. As in the bunk cars, all of which found their model in the sleeping arrangements of the canal boat, the berths were only on one side of the car and

[1] *New York Commercial Advertiser*, Nov. 30, 1875.

consisted of a triple tier of two double and one single berth; an arrangement later changed to one double and two single berths.

The Woodruff sleeping car also was designed about this time by T. T. Woodruff, Master Car Builder of the Terre Haute & Alton Railroad. In this car both sides of the car were utilized as in the Pullman car, and the sleeping accommodations consisted of twelve sections, six on a side. A company was formed to operate the Woodruff cars in 1871, with a capital of $100,000.

The Flower Sleeping Car Company was another characteristic competitor. This short-lived company was organized in 1882 in Bangor, Maine, with a capital of $500,000. The seats in this new car were placed in the middle instead of on the sides of the cars, thus leaving an aisle on each side instead of one in the center. Claims were made that a freer circulation of air would result, and a news item of the *Times* further recommended this unique construction as more convenient to families, the berths being so arranged, side by side, that two could be made up into a double bed.

Mann's Boudoir Car Company was incorporated

in 1883, with a capital of $1,000,000, and experienced considerable popularity due to their unique arrangement, which has been described in a previous chapter.

In 1883 the Erie Railroad realized the long entertained ambition of entering Chicago on its own rails. To accomplish this, the Erie had leased the New York, Pennsylvania & Ohio Railroad and built the Chicago & Atlantic. Through connection was actually made May 15, on which date freight traffic was begun.

The train by which the Erie inaugurated the passenger business over the new trunk line was probably the most complete and elegant train ever to that time constructed. All of the cars were of Pullman manufacture and consisted of a baggage car, second-class coach, a smoking car, and first-class coaches and sleepers that were "models of perfection and beauty, as might be expected where the Pullman Company had *carte blanche* to produce the best possible." Each coach was lighted with the new Pintsch lights. The smoking car deserves more than passing mention, for it was the first one ever constructed of Pullman standard. The car was equipped with upholstered easy chairs, and a "refreshment buffet" moistened the throats of the smokers.

Early in 1889 the Pullman Company acquired the control of the Mann Boudoir Car Company and the Woodruff Sleeping Car Company, including the entire car equipment and plants. By this acquisition a long step was taken for the unification of sleeping car service, and the further development of a uniform and widely extended scope of operations. For years the success of the Pullman Company's service had been too generally acknowledged to escape the notice of enterprising railroad men, and these two companies were fair examples of the numerous competing companies that were organized. But the success of the Pullman service was based on an idea of too wide conception ever to be successfully imitated. The success of the company engendered competition; its success resulted only in a comparison of service injurious to the imitators. Behind all this lay the fundamental reason for Pullman supremacy. Created to give a standardized service everywhere for the convenience of travelers, it was quickly apparent that competition was but a reversal to the old order — the more companies, the less uniform service.

About a month previous, the Mann Boudoir Company and the Woodruff Sleeping Car Company had joined hands and formed the Union Palace Car Company. By the purchase of this combine the Pullman Company added about 15,000 miles of road to that already operated, and by that many miles extended its through car service. The only remaining sleeping car companies of any importance outside of the Pullman Company were the Wagner Company, belonging to the Vanderbilts, and operated over the Vanderbilt lines, and the Monarch Sleeping Car Company, which operated entirely in the New England States with the exception of one Ohio line. A newspaper of the time commented on the merger, and closed with the verdict: "While this will add to the volume of the Pullman business, it will also render the service upon the absorbed lines far more efficient and satisfactory for the traveling public."

In 1888, Mr. Pullman had put in operation his vestibule trains, which immediately met with extraordinary favor and patronage. In a very few days the Wagner Company also advertised a vestibule train and were promptly met with an injunction holding the Wagner appliances to be an infringe-

The first step in the building of the car. The center construction in position, and the framework assembled

ment of the Pullman patent. After another hearing, the injunction was superseded, the Wagner Company giving an unlimited bond, signed by the Vanderbilts, to pay any damages ascertained by the courts.

After months occupied in taking the evidence of travelers, expert mechanics, railroad officials, prominent citizens, and others, a final hearing was had. The judges, owing to the vast interests involved and the legal difficulties presented, took ample time for consideration, but finally adhered to their first conclusion. The main feature of the Pullman vestibule system was the Sessions patent, without which the vestibule system was worthless. The court declared this invention to be of the highest order of utility, not only as shown by the testimony in the case and the adoption of the patent by the principal railroads of the country, but also by the acts of the Wagner Company in appropriating the device, and in the tenacity with which they clung to it in the courts under an immense bond for any damages to result, and so, in April, 1889, the United States Circuit Court delivered its opinion in favor of the Pullman Palace Car Company in its long and stubborn fight with the Wagner Palace Car Company.

VI

The Town of Pullman

CHAPTER VI

THE TOWN OF PULLMAN

LIKE most other industries, the Pullman Palace Car Company felt the effect of the financial depression immediately following 1873, but the reaction followed, and on the resumption of specie payments in 1879 dawned a new era in the Company's history and a rapid expansion of its business. To meet this expansion and to extend the business still farther along the line of general car building, it became necessary to enlarge the plant. The shops already established in St. Louis, Detroit, Elmira, and Wilmington were unable to provide the volume required by the increasing demand for the Company's output. It was evident that new shops must be built on a larger and more comprehensive scale than any that had gone before.

In 1879 the Chicago newspapers were alert to confirm the rumor that George M. Pullman was planning to locate his new shops at Chicago. The following year the rumor became fact and the question of the exact location became of paramount interest.

Chicago with its central position with reference to the railway systems of the continent, seemed the natural site, but there were weighty objections, touching both finance and the matter of labor, to be urged against building within the city limits proper. Sites were visited by representatives of the Company at Hinsdale, Illinois, and Wolf Lake, Indiana, but in April it was definitely announced that the works would be located on the Illinois Central Railroad on the shore of Lake Calumet. A Chicago newspaper commented on the decision of the Company as follows:

> A notable addition to Chicago's mercantile industry is to be the extensive car works of the Pullman Palace Car Company, ground for which is to be broken today. A larger establishment for manufacturing purposes will not exist in the West, and while it will contain all the latest and most improved mechanical appliances in use, it will embody in its architecture grace and beauty that is quite characteristic of the palace car. The works are to cost $1,000,000; about 2,000 men are to be employed in them, and the extended arrangement of machinery is to be moved by the Corliss engine, one of the Centennial wonders, which has been purchased by the Pullmans.

Fitting the car with steam pipes and electric conduits

At work on the steel plates for inside finish panels

THE TOWN OF PULLMAN

An interesting personal reminiscence of this famous real estate operation may be found in Frederick Francis Cook's *Bygone Days in Chicago*.

Another "Pullman scoop" was of an extraordinary real-estate and manufacturing interest when "negotiated"—the slang to be accepted for once in its proper meaning. In the later seventies, besides other duties, I had charge of the real-estate department of the *Times*. It became known that the Pullman Company intended to build a manufacturing town somewhere, but whether in the environs of Chicago, St. Louis, Kansas City, or other western point, was for the public an open question for many months—and, I dare say, for a time was an unsettled proposition with the company itself, for St. Louis offered large inducements in the way of land grants. What finally turned the scales in favor of Chicago, according to Mr. Pullman's declaration to me, was the more favorable climatic conditions presented by Chicago. It was his contention that during the summer a man could do at least ten per cent more work near Lake Michigan than in the Mississippi Valley in the latitude of St. Louis.

During many disturbing weeks—for the whole real-estate market in at least three cities waited on the decision—frequent announcements were made that the directors of the company, or its committee on site, had inspected this locality, or that, in the vicinity of one city or another, and so the wearisome time went on. Many places were visited about Chicago—some to the north, some on the Desplaines, some in the neighborhood of the

Canal, but somehow none near Calumet Lake, a fact which finally aroused my suspicions. In the meantime, unverifiable reports of large transactions in that locality floated about in real-estate circles. Finally, I pinned down an actual sale of large dimensions, with Colonel "Jim" Bowen as the ostensible purchaser. That opened my eyes, for the colonel's circumstances at this time put such a transaction on his own account altogether out of the question.

Almost daily at this time Mr. Pullman was interviewed on the situation by the real-estate newspaper phalanx—Henry D. Lloyd was then in charge for the *Tribune*—but "nothing decided," was the stereotyped reply. By and by I discovered that almost invariably if I went at a certain hour, "Colonel Jim" would be largely in evidence about the Pullman headquarters, with an air of doing a "land-office business," and, as it turned out, he was actually doing something very much like it. Slowly I picked up clue after clue, pieced this to that, and one day felt in a position to say to Mr. Pullman that I had located the site. He seemed amused, and laughingly replied that he was pleased to hear it, as it would save the committee on site a lot of trouble; and, as some of them were that very day looking at a Desplaines River site near Riverside—a trip most ostentatiously advertised in advance—he thought he would telegraph them to stop looking, and come back to town.

It was always a pleasure to interview Mr. Pullman, for he had a way of making you feel at ease, and I entered heartily into the humor of his jocularity. But, as in a bantering way, I let out link after link of my chain of evidence, he became more and more serious, and finally—

without committing himself, however—took the ground that even if true, in view of the importance of their plans, no paper having the good of Chicago at heart ought by premature publication to interfere with them. He pressed this point more and more, and finally made frank confession that I was on the right track, by acknowledging that they had already bought many hundreds of acres, were negotiating for many hundreds more which would be advanced to prohibitive prices by publication, and the whole scheme would thus be wrecked. On the other hand, if I withheld publication, he promised that I should have the matter exclusively—the whole vast improvement scheme, unique plan of administration, etc. As there was the danger in waiting that one of my rivals might get hold of the facts, exploit them, and thus turn the tables on me, I replied that the matter was of too great moment for me to take the responsibility of holding the news, and that I should have to consult Mr. Storey. It happened that Mr. Storey had invested quite extensively in South Side boulevard property; and, as a great improvement southward could not fail to add to the value of his holding, and there was the further prospect of a more complete exclusive account later than was possible with my skeleton information, he gave a ready assent.

The town of Pullman meant far more in the mind of its founder than a mere industrial establishment. The dreary, water-soaked prairie was raised to high, dry land; an entire town was planned and blocked out following Mr. Pullman's own design. Architects and landscape architects worked together to carry out the plan to a harmonious and pleasing fulfillment. Among the more prominent details of this vast work were included a system by which the sewage of the town was collected and pumped far away to the Pullman produce farm; the equipment of every house and flat regardless of rental with the most modern appliances of water, gas, and plumbing; the establishment of athletic fields; the concentration of the merchandising of the town under the glass roof of the central arcade building, and the construction of a handsome market house, a fine schoolhouse to accommodate a thousand pupils, a library containing over 8,000 volumes, a savings bank and a large and artistically decorated theater. The population of Pullman in January, 1881, counted four souls. In February, 1882, there were 2,084 inhabitants, a total which had increased to 8,203 by September, 1884.

A contemporary writer closes an enthusiastic description of the town of Pullman with the following paragraph:

> Imagine a perfectly equipped town of 12,000 inhabitants, built out from one central thought to a beautiful

Preparing the steel frame for the upper section of a Pullman sleeping car

Sand blasting the brass trimmings of the car before applying the finish

and harmonious whole. A town that is bordered with bright beds of flowers and green velvety stretches of lawn; that is shaded with trees and dotted with parks and pretty water vistas, and glimpses here and there of artistic sweeps of landscape gardening; a town where the homes, even to the most modest, are bright and wholesome and filled with pure air and light; a town, in a word, where all that is ugly, and discordant, and demoralizing, is eliminated, and all that inspires to self-respect, to thrift and to cleanliness of person and of thought is generously provided. Imagine all this, and try to picture the empty, sodden morass out of which this beautiful vision was reared, and you will then have some idea of the splendid work, in its physical aspects at least, which the far-reaching plan of Mr. Pullman has wrought.[1]

[1] *The Story of Pullman*, prepared for distribution at the World's Fair, 1893.

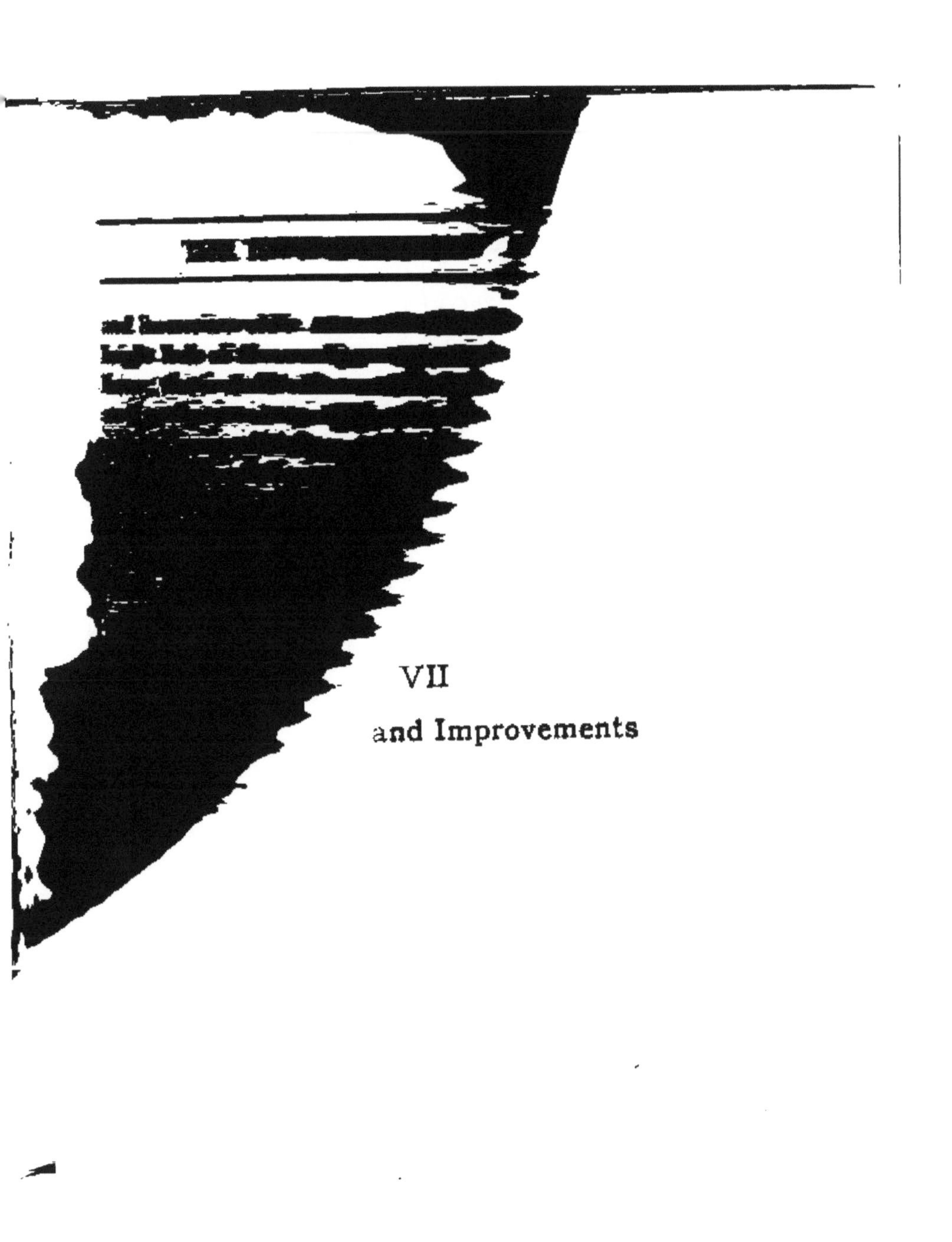

VII

and Improvements

VII

Inventions and Improvements

CHAPTER VII

INVENTIONS AND IMPROVEMENTS

THE invention of the folding upper berth combination by Mr. Pullman was the first of many contributions by himself, and in later years by the Pullman Company and those associated with it, to the development of railway travel. Sleeping cars for a number of years had given night accommodations to travelers; there was nothing new in the idea that a night journey required sleeping accommodations. But in the new and radical berth construction devised by Mr. Pullman lay the difference between impracticability and practicability—between discomfort and luxury.

The earliest sleeping cars were mere bunk cars in which the male passengers might recline during the night hours. Later, bedding was furnished, but the necessity of storing it by day in a closet at the end of the cars created a situation in which order and cleanliness were far from practicable. By the Pullman invention, however, all this was changed. A type of car was developed that was not only comfortable and convenient for day travel, but one that might be quickly transformed into a comfortable sleeping apartment. Furthermore, the new upper berth construction made it possible to pack away by day the entire bedding, mattresses, curtains, and partitions necessary to convert each section into a double sleeping apartment.

With this simple mechanical innovation the inventor combined an idea characterized by a breadth of vision that ranks with the great ideas of the century. In few words, he conceived the thought that it would be possible at one stroke to supplant the inadequate and inefficient service of the day with a new service so complete in its comforts and conveniences that no one might express a wish that the service might be unable to fulfill.

It is interesting, in passing, to consider the fact that up to the development of the Pullman car, night trains were patronized exclusively by men, for no woman would have considered subjecting herself to the inconvenience and lack of privacy of the ordinary sleeping car. The development of the Pullman car and Pullman service made continuous day

View of machine section. Steel Erecting Shops

Fitting up the steel car underframe. Steel Erecting Shops

and night travel practical for women and children; it created the comforts and privacies they naturally required. To be sure it was several years before the new order of things received general recognition, but the public quickly caught on. "Travel by Pullman" soon became a popular diversion.

The story of the early years of the Pullman sleeping car has been told in the foregoing chapters. Due in large measure to the comfort and convenience of the cars, continuous travel lengthened, and at once arose the necessity for eating as well as sleeping accommodations on the through long-distance trains.

For a number of years foreign travelers in America had praised the elaborate restaurant service afforded by certain station eating-houses. Towns developed keen rivalry in respect to the meals provided by their station "counters," and the station restaurants of certain towns developed among constant travelers a reputation for unusual culinary excellence. Our fathers will doubtless recall the glorious fame of dining rooms at Poughkeepsie, Springfield, and Altoona, and of certain dishes that enjoyed nation-wide reputation and might be had only at this or that particular station restaurant.

But, on the other hand, the uninviting, indigestible nature of the so-called refreshment offered at some railway eating stations had long been a byword. In most sections of the country it was practically impossible to procure a respectable meal or lunch while traveling. Railway officials had wrestled with the subject in vain. Recognizing the fact that the heart of the railway traveler is most susceptible to influences reaching it by way of his stomach, they made repeated and continued endeavors to improve the fare offered during the "twenty minutes for dinner" stops. With a few exceptions the results were not encouraging, and the traveling public continued its dyspeptic round three times a day.

The station eating-house was on an unsound basis, and its disadvantages were obvious. With the increase of the speed of through trains and the demand for shorter running times between terminals it became quickly apparent that a train could not be stopped three times a day to permit the passengers to gorge a hasty meal at the station restaurant. Three meals at a minimum of twenty minutes each was an hour lost, and twenty minutes for eating was as bad for the passenger as it was for the running time of the trains. There were still other disadvantages. In addition to the delay of the train and the tax on the passenger's digestion, there was the frequent discomfort of wet or wintry weather. On a fine day it was well enough to "stretch one's legs," but in rain or snow the tri-daily evacuation of the car was a decidedly unpopular feature.

The installation of "hotel-car" service by the Pullman Company sang the knell of the station eating-counter. The "President," a car combining sleeping and eating accommodations, was put in service in 1867 on the Grand Trunk Railway, then the Great Western of Canada. Its instant success necessitated the building of the "Kalamazoo" and "Western World," and in the years immediately following many hotel cars were put in service.

The second step in the evolution was inevitable. At best, the hotel car was only a sleeping car with restaurant accommodations. Eating and sleeping have never been associated in the modern mind; there must be a separate place for each.

To meet the demand, or rather to anticipate a demand which his keen eyes foresaw, Mr. Pullman set himself to the task of developing a car which would be only a dining car, serving no other purpose, and practical for operation in conjunction with through trains of the fastest speed. The first real dining car which Mr. Pullman constructed was aptly named the "Delmonico." It was a complete restaurant with a large kitchen and pantries at one end. The main body of the car was fitted up as a dining room in which the passengers from all the cars of the train could enter and take their meals with entire comfort. The "Delmonico" was put in regular service in 1868 on the Chicago & Alton, and other Pullman diners were added the same year. At about the same time the Michigan Central and the Chicago, Burlington & Quincy Railroads also began to operate dining cars on their trains. To the Chicago & Alton, however, belongs the honor of having first inaugurated the dining-car system. The Michigan Central and Burlington did not put on dining cars until 1875. The Chicago & Alton dining cars were run between Chicago and St. Louis, and were constructed and managed by Mr. Pullman. The price for a meal was $1.00. Later the Alton acquired an interest in the dining cars, and finally assumed full control of them.

Making the cushions for the seats. Upholstery Department

Making the chairs for the parlor cars. Upholstery Department

Although founded and developed, and for a number of years successfully operated by the Pullman Company, the dining car is no longer under its management. Due primarily to the vast increase in this particular share of the business and the variety of service required by travelers in different sections of the country, it became advisable to turn over to the various roads the details of catering to their particular patrons. On some of the leading railroads the highest type of dining-car service is maintained and advertised as a particular feature. On other roads of lesser prominence a corresponding degree of service may be found. It is, perhaps, unfortunate from the point of view of the traveler that the Pullman Company found it necessary to discontinue a service that it had so auspiciously inaugurated.

The installation of dining-car service immediately drew attention to a serious defect in railway train construction that had previously escaped notice, a defect which was the more apparent in comparison with the relatively high development of other features of train construction. By the adoption of the dining car it became necessary for the passengers to pass from car to car across the platform while the train was in motion, and often during a condition of rain and snow which added discomfort to actual danger. Where the crossing of platforms while the train was in motion had formerly been prohibited, the railroads were now forced to encourage passengers to subject themselves to this dangerous procedure in order that they might avail themselves of the convenience of the dining cars.

Attempts had been made at different times to provide a safe and covered passageway between the cars, especially on fast express trains, but nothing of a practical nature had resulted. In 1852 and 1855 patents were taken out for canvas devices to connect adjoining cars and create a passage way between them. These appliances were installed in 1857 on a train on the Naugatuck Railroad, in Connecticut, but soon proved to be of little practical use and were abandoned several years later.

But in 1886 Mr. Pullman, realizing the handicap of existing conditions to the full enjoyment of the various types of cars which he had established, set himself to the solving of the problem by devising a perfect system for constructing continuous trains and at the same time providing sufficient flexibility in the

The frame end posts for Pullman standard cars are made in this section of the shops

The assembling of the steel car partitions is shown in this picture

connecting passage ways to allow for the motion of the train, particularly when rounding curves. The result of his efforts combined with those of his associates was the complete solution of the problem and the establishment of the "vestibule" train, practically as it exists today. The vestibule patent was granted to Mr. H. H. Sessions, of the Pullman Company, and covered many important features, and particularly the arrangement of the springs which kept the cars in line in a vertical plane.

The vestibule was patented in 1887. By its application the appearance of the train as a unit was materially increased, but of far greater importance was the contribution which it made to safety. Not only did the enclosed vestibule afford protection to passengers crossing the platform from one car to another, but the entire vestibule construction immediately gave greater safety in case of wreck by preventing one platform from "riding" the other and producing a telescoping of the cars.

The vestibule as designed and patented did not extend to the full width of the car. It consisted of elastic diaphragms on steel frames attached to the ends of the cars, the faces of the diaphragms when the train was made up, pressing firmly against each other by powerful spiral springs which held them in position. A further advantage of the vestibule was

The vestibule was invented by George M. Pullman. This illustration shows its earliest form which extended only to the width of the doorway of the car. In 1893 it was extended to the full width of the car.

the almost entire elimination of the oscillation of the cars.

The first vestibuled trains were put in service in April, 1887, on the Pennsylvania Railroad, and in a few years were adopted by every railroad using Pullman equipment. In 1893 the vestibule was redesigned to enclose the entire platform by means of a drop which lowered over the stair openings, thus increasing the roominess of the car and utilizing every inch of possible space.

In the *Railway Review* of April 16, 1887, occurs an interesting description of the first "solid-vestibuled" train. For a number of months following, this radical innovation was widely recognized by the press throughout the country, and Pullman vestibuled cars were advertised by the railroads on which they were operated. We quote in part from the article in the *Railway Review:*

This week there was turned out of the Pullman works, at Pullman, Ill., a train of three sleepers, one dining car, and one combination baggage and smoker, that for perfection, in detail of manufacture and ornament, and in completeness of comfort and luxury, is unquestionably far ahead of any train ever before made up. This train was on public exhibition for a few days at Chicago, and on

Friday was taken on its christening trip, over a short run on the Illinois Central Railroad. The train is intended for "Limited" service on the Pennsylvania system.

The trial trip was a success in every way. The train went to Otto, a short distance south of Kankakee, sixty miles from Chicago. There it was reversed on a Y, and an opportunity afforded of witnessing its operation on a sharp curve. The action of the flexible connection of the vestibules was perfect. On the return trip the train was run at a high rate of speed, and it was evident that the cars were held very firmly together, by the springs at the top of the vestibules, and that there was much less jarring and swaying than is usual even on a very level track.

The list of business men and railroad managers who made up the party indicates the importance of the occasion. It included:

George M. Pullman
G. F. Brown
T. H. Wickes
C. H. Chappell
J. J. Janes
Orson Smith
O. W. Potter
W. T. Baker
H. R. Hobart
A. N. Eddy
Jesse Spalding
Frederick Broughton
W. P. Nixon
John M. Clark
A. C. Bartlett
J. W. Hambleton
E. L. Brewster
Henry S. Boutell
D. B. Fiske
Willard A. Smith
Stephen F. Gale
Edson Keith
O. S. A. Sprague
A. B. Pullman
J. T. Lester
H. J. MacFarland
S. W. Doane
Murray Nelson
A. H. Burley
C. K. Offield
E. T. Jeffery
Prof. Swing
W. K. Sullivan
W. K. Ackerman
A. C. Thomas
J. McGregor Adams
J. F. Studebaker
P. E. Studebaker
T. B. Blackstone

Axle generator for electric lighting of the car

Rev. S. J. McPherson, A. A. Sprague, D. S. Wegg, P. L. Yoe, F. N. Finney, C. S. Tuckerman, A. F. Seeberger

During the days in which the train was exhibited at Van Buren street, Chicago, it was visited by approximately 20,000 people. The article continues:

> This fact shows that the public has a deep interest in improvements in traveling conveniences. We do not remember that any previous invention or improvement has ever excited such general public interest. Mr. Pullman has again struck the popular chord.

The first vestibule train to the land of the Aztecs, the "Montezuma Special," was naturally of Pullman construction, and began regular tri-monthly trips from New Orleans to the City of Mexico and return, via the Southern Pacific, Mexican International, and Mexican Central Railway, on February 7, 1889. Four magnificent cars, electrically lighted, comprised the train. The initial trip of 1,835 miles was made in about seventy-one hours, and on its arrival in the City of Mexico a banquet was given to President Diaz and his cabinet to signalize the advent of the first international vestibule train into the capital of Mexico.

The lighting of railway cars shows an interesting evolution. Undoubtedly candles were used at the earliest period, but the use of oil dates back beyond the birthday of the Pullman car. Oil lamps, at best, were a poor substitute for the light of day. Casting a dim, yellow light, flickering in every draught, smelling and smoking when not properly cared for, and vitiating the car atmosphere, it was small wonder that the public showed prompt appreciation of the first substitute that was provided.

The brilliant Pintsch light, which for a number of years had had wide use in Europe, was first introduced into America by the Pullman Company on the crack Erie train in the through New York-Chicago service in 1883. The gas used for these lights was of high candle power and was manufactured from petroleum. As a car illuminant it has held its own almost to the present day.

It is impossible to exaggerate the part played by the Pullman Company in the development of electric lighting of cars. Without its inspired initiative and its vast resources for practical and costly experiment it is fair to believe that electricity would not have been successfully utilized for this purpose for many years. The *Railroad Gazette* of January 25, 1889, expresses this thought:

> Without extended experiments we can scarcely hope to develop a good system of electric lighting for railroad service. Such experiments are rather expensive, and it is only by the co-operation of liberal-minded managers that anything like a perfect sy‹ .em can be expected in a reasonable time. The Pullman Company has great confidence in the success of electric lighting, and therefore, in spite of the annoyance and expense of the present system, expresses a determination to use it, expecting that something better will result in the near future from the extended experience now being obtained.

Although the incandescent electric lamp was introduced by Edison in 1879, following by two years the introduction by Brush of the arc lamp, it was on an English railway in an American Pullman car supplied with electricity by French accumulator cells that the electric light on October 14, 1881, barely fifty years from the first suggestion of the iron horse by Stephenson, cast its brilliant light for the first time in a railway carriage.

The trial was made in a Pullman car, forming part of a special train on the Brighton Railway. A number of officials of the road, a representative of the Pullman Company, and Mr. F. A. Pincaffs and

Mr. Lachlan of the Faure Accumulator Company composed the party, and at 3:25 the train pulled out of the Victoria Station for Brighton.

Only a few months before, Mr. Faure had sent to Sir William Thomson his little box of lead plates coated with red oxide and fully charged with electricity. The great physicist saw at once its possibilities, and in a relatively short time inventors were developing countless applications of the new wonder. Its application to car lighting was an important test.

The Pullman car on which this first experiment was made, carried beneath it on a shelf some thirty-two small metal boxes or cells, each containing lead plates coated with oxide. Stored in these cells was the power to light the car. It was nothing more than the most elementary storage battery, a far cry from the compact batteries of today and the massive generator swung beneath the floor of the modern car.

All the previous night a steam engine had created power to charge the cells. In the roof of the car were twelve small Edison incandescent lights with bamboo filaments. The light was uneven; it was "garish," but at the turn of a switch its rays filled the car. With pardonable enthusiasm the *London*

The sewing room. Upholstery Department

Times stated that "the car on the return journey in the evening was kept lighted the whole of the distance from Brighton to Victoria."

It is interesting to read in the *London Daily Telegraph* of October 15, 1885, the following mention of this important event:

Yesterday's trial was understood to have special reference, however, to a new train, wholly composed of Pullman cars, which it is proposed shortly to put on the service between Victoria and Brighton, and should the experiment be deemed fully satisfactory it is probable that the new train will from the first be fitted with the electric light. So far as the travelers were concerned the result was eminently successful. It would scarcely be possible to conceive a steadier, more equable, or more agreeable light. On the down journey the first trial was made in the Merstham tunnel, and then in the Balcombe and Clayton tunnels. All that was needed was to move the little switch, and instantaneously the delicate carbon thread enclosed in the lamps was aglow with pure white light. The return journey was made in the night, and the electric lamps were alight during the whole distance. There had been some question whether the supply would prove sufficient, as owing to stoppages the special had taken a somewhat longer time than had been allowed for; the event, however, showed that the storage had been ample. It would be possible to generate electricity by the energy of the moving train itself, and this has indeed been suggested to be done. By this means enough energy could be supplied to the incandescent lamps, but in any case the accumulator would be necessary to act as a reservoir when the train was not in motion. It possesses, however, another advantage equally important. Experience shows that a current of absolutely uniform strength supplying an even and constant light can only be derived from stored electricity. The oxide of lead which covers the plates not only prevents leakage, but enables the supply to be withdrawn with perfect regularity, and renders sub-division easy. Yesterday the smoke room and lavatory of the car were lighted, and occasionally the lights were turned off without in any way interfering with the other lamps in the same circuit. Before the train started on the return journey the brightly illuminated carriage was an object of interest to many members of the Iron and Steel Institute who visited Brighton and Newhaven yesterday. With regard to expense, it is claimed for the accumulator and the incandescent lamps that the expenditure would be decidedly less than on oil, while, as to the comparative value of the two there is no room for difference of opinion. It was the general feeling of all who took part in the excursion that the question of the electric lighting of trains had been solved, and that to the Brighton Company, whatever may be the immediate results of the experiment, would belong the honour of taking the first decisive and practical step in the way of reform.

Four months later a correspondent of a Sheffield, England, paper, writing from London to the *Railway Review* of the recent trial of electric lights on the Pullman train of the London, Brighton & South Coast Railway, says:

There is no doubt whatever on the point that this, apart from the question of cost, is a decided success. It is easily manageable, and diffuses through the train a pleasant, equable light, scarcely less agreeable than daylight. It is turned on and off with instantaneous effect as the train enters and leaves a tunnel, and of course is kept burning the whole of the time during the night journeys. The electricity is stored in a number of lead plates, which are kept in water in iron boxes in the guard's van. There are two lots, one at either end of the train, and two mechanics in charge of them. This discovery of the ability to store electricity for application to lighting purposes seems to carry the discovery farther than anything since it was first introduced. It gets over many difficulties which seemed insuperable—especially the important one of the great waste of power which is illustrated every night at the Savoy Theatre—and would be applicable to the introduction of electricity for household use.

At the Savoy, when the exigencies of the play require that the lights should be turned down in the auditorium, there is no cessation of the enormous power required to produce the full effect. What happens is that by a mechanical contrivance, the electricity is carried off from the light and goes to waste. With this system of storing, electricity can be used just like gas, as much or as little as people chance to want. Another great advantage is the freedom from jumping, inseparable from the action of the driving power of the steam engine, or of the motion power of water. The lights of the Brighton train burn just as steadily as gas, an effect not in any way obtained where the light is maintained directly by the driving power of steam.

But after all, the question of gas vs. electricity will resolve itself into one of cost, and it is here where gas will inevitably hold its own. The fundamental principle of the electric light is that for a given exertion of power you obtain a given proportion of light, neither more nor less. For every hour it is burning there will be required a certain exactly-ascertained proportion of revolutions of the steam engine, and therefore, if the whole town is lighted it can be done only at a strictly proportionate expense to the lighting of a single house. As to what that expense will be, as compared with gas, the Brighton train would, if we had an idea of the actual figures, afford a precise means of information. I met on the train a well-known gas engineer, attracted, like myself, by the novelty of the experiment. What the electric light cost he was not able to say, but when we take into account the capital sunk in plant, involving a steam engine with the necessary buildings, consumption of coal and necessary employment of skilled labor, it must be something considerable. Against this is the bare fact that the Brighton train could be lighted with gas for the double journey at the cost of 10d. It is a physical impossibility that electricity should ever come anywhere near this, and that probably explains the singular phenomenon that at the time when electricity is making conspicuous advances in public favor, the value of gas shares is not only steadily maintained, but is actually rising in the market.

The steel parts used for interior car finish are all standardized, and are formed by powerful presses

Another large press at work on the forming of steel shapes for the interior framing of the cars

INVENTIONS AND IMPROVEMENTS

The present method of heating an entire train with steam from the locomotive was satisfactorily tested out in the winter of 1887, and was generally adopted the following year. By this improved system the individual heaters in each car were abolished, and a source of much discomfort and complaint was removed. The Pullman cars were immediately altered to benefit by the new system.

VIII

How the Cars are Made

CHAPTER VIII

HOW THE CARS ARE MADE

IN former chapters has been told the story of the birth of the Pullman car and its development through the various phases of its evolution. Generally speaking, this evolution for the first forty years was characterized chiefly by the addition, at one time or another, of certain inventions and improvements, such as the electric light and the vestibule, and by a changing style of interior decoration conforming to contemporary fashions. But at no time is recorded a change in the basic idea of car construction that can in any measure compare with the revolutionizing change which was recorded in 1908 by the construction of the first "all-steel" Pullman car.

For a number of years steel sills and under frames had furnished a staunch foundation for all cars manufactured by the Pullman Company for its operation. Further strengthened by steel vestibules, it is to be doubted if the all-steel car offered any very material increase in the safety already afforded to the passengers. But the change which the steel car brought in the process of manufacture was radical in the extreme. The first Pullman cars, and in fact every car up to and through the nineties, was of all-wood construction. Wood-making machinery filled the great shops at Pullman; carpenters and cabinet-makers numbered a big percentage of the pay roll. It was a wood-working industry. At one fell stroke the old order changed to the new. The songs of the band-saw and the planer were stilled and in their stead rose the metallic clamor of steam hammer and turret lathe, and the endless staccato reverberation of an army of riveters. Ponderous machines to bend, twist, or cut a bar or sheet of steel filled the vast workrooms. An army of steel workers, Titans of the past reborn to fulfill a modern destiny, fanned the flames in their furnaces and released the leash of sand blast, air hose, and gas flame.

But fascinating as unquestionably was the work of the patient artisans who inlaid the beflowered Eastlake Pullman or the Moorish cars of another day, there is equal romance in the product of the modern worker who builds these rolling hostelries

This machine is at work punching holes for screws etc. in the steel for the inside finish

This great power press is engaged in shaping the steel panelling for the inside finish of the car

of steel. Under the high glass roof the tumult of ponderous machines fills the air with pandemonium. At one side of one of the main aisles a half dozen great steel girders, like keels for giant ships, lie on the floor. These are the mighty box girders, eighty-one feet in length and weighing over nine tons each, which will form the backbone of future Pullmans. To each of these girders, or sills, are riveted plates, angles, and steel castings which extend the full length of the car and platforms, as well as floor beams, cross bearers, bolsters, and end sills of pressed steel. On this foundation the side sills are riveted, steel beams that run the entire length of the car.

When this gray mass of steel is finally riveted together with its coverplates, tieplates, and floor-plates, the underframe of the car is completed—an almost indestructible foundation which alone weighs 27,365 pounds. On this underframe the superstructure or frame is erected to form the body of the car. This frame is composed of pressed steel posts and plates forming for each side a complete girder which would by itself alone carry the entire weight of the loaded car.

The roof deck is separately assembled, and as soon

as the superstructure of the car is ready it is swung up by a crane and dropped into place. Like the rest of the car, the roof is of steel, braced and riveted to defy the greatest possible strains. The ends and vestibules are now built on, piece by piece, until the skeleton of the car is complete. The vestibules are particularly imposing, for on each side, framing the side doors through which the passengers enter the car, are giant beams of steel so built into the construction of the frame that only under most extraordinary circumstances could the force of a collision crush the vestibule or the car behind it.

The trucks which carry this tremendous burden of steel are marvels of strength and efficiency. Each of the two trucks has six steel wheels weighing nine hundred pounds apiece. Added to this is the weight of the three six hundred pound axles, the two steel castings which form the framework for the trucks together with the bolsters, springs, equalizers, and brake equipment — a total weight of 42,000 pounds for the trucks alone, contributed to the total weight of the car.

The car is now subjected to a thorough sand-blasting, a process that removes every particle of

Riveting the underframe

The steel end posts in position, providing strongest possible protection in case of collision

scale, grease, or dirt and leaves the steel in perfect condition to receive the first coat of paint and the insulation. To the passenger, the presence of the steel construction is apparent, but the insulation, which forms a vital factor in the car's construction, can be seen only during the process of building. Composed of a combination of cement, hair, and asbestos, this insulating material is packed into every cubic inch of space between the inner and outer shells of the roof and sides, forming a perfect non-conductor to protect the passengers against the biting cold of winter or the heat of summer sunshine. A similar cement preparation is next laid on the floor, combining the quality of a non-conductor of heat and cold with sanitary qualities invaluable as an aid in maintaining the cars in a strictly sanitary condition.

At this point in the construction the car is turned over to the steamfitters, plumbers, and electricians, who perform their work with the skill and dispatch bred of a long familiarity with the particular requirements of car construction. To see the Pullman car at this stage is to see a network of steam-pipes and electric conduit lacing in and out between the gaunt steel frame of the car, and everywhere the white plaster-like insulation packed into every cavity. As soon as these gangs of workmen have finished, other workers fit into place the interior panel plates, partitions, lockers, and seat frames, and the car instantly assumes a new and almost completed aspect. Meanwhile the painters have completed their work on the exterior of the car and begin the finer finish of the interior. Here coat upon coat is laid, and after each coat laborious rubbing to give the required finish. The graining, by which various woods are so faithfully imitated, is then applied, and last the varnishing.

The car is now completed with the exception of the fittings. A gang of men hang curtains in the doors and windows; the upholsterers contribute the carpets, cushions, mattresses, and blankets; the various little fixtures are added, and the car is finished. *Steel! Veritably!* One man can trundle in a single wheelbarrow all the wood that has gone into its construction.

Rich Brewster green, the new paint gleaming in the sunlight, a long line of these seventy-ton steel mile-a-minute hostelries are waiting for the hour

Type of wood-frame truck used on early cars; four wheels only, with a big rubber block over each in place of springs

Modern cast-steel truck; six wheels with powerful springs to take up the jars and jolts of the road

when the white-jacketed porters will open their doors in welcome to their first passengers. Above the windows the word "Pullman" in dull gold will carrry from coast to coast the name of their founder. Below the windows is the name of the car, selected usually with local significance in consideration of the lines over which that particular car will operate.

In a corner of the great yards at a track end stands a little yellow car, smaller than many of our interurban trolley cars, the paint peeling from the boards that have seen the changing seasons of half a century. It is old number "9," not the earliest, but one of the early Pullmans. Perhaps there are nights, when the roar of the machines is stilled, that the ghosts of a long-past day once again walk up and down the narrow aisles, strangers to the age of steel.

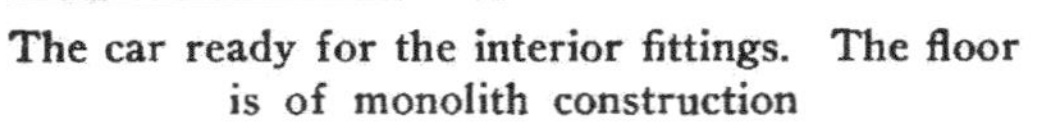

The car ready for the interior fittings. The floor is of monolith construction

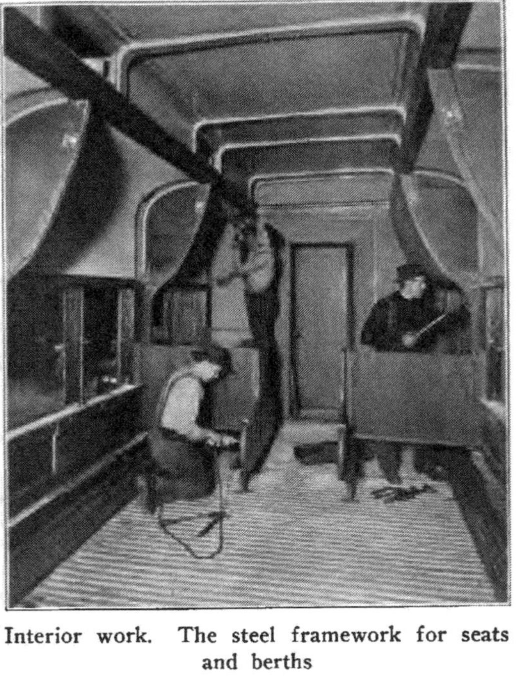

Interior work. The steel framework for seats and berths

IX

The Operation of the Pullman Car

CHAPTER IX

THE OPERATION OF THE PULLMAN CAR

ON the magic carpet of Bagdad the fortunate travelers of a fabulous age were transported to their destination, over valley, river, and mountain with a certainty and dispatch that has been unparalleled in the annals of passenger transportation. But the magic carpet, despite the generous measure of its service, seems to have been lost to following generations, and only its reputation, doubtless somewhat amplified by the telling, remains to set a high standard to succeeding transportation enterprises.

Service is a much-used and a much-abused word. It has manifold significance. It may be a personal thing and carry the conscientious effort of individuals eager to do for others offices which they desire performed; it may be purely mechanical and consist only in the provision of the "ways and means" to secure a desired end. It may be a combination of both; a system or organization instituted for the accomplishment of a duty or work beneficial to a community. A great railroad affords such a service. Greater in its scope than any railroad, the Pullman Company provides a more vast, intricate, and complete service to the people of the United States, a service unequaled in all the world.

A study of the scope and ramifications of the Pullman operations deserves more than passing comment; it is of interest to everyone, for everyone is to some degree a traveler; an actual or a potential Pullman patron. In preceding chapters has been traced the story of passenger transportation in America; how the first railroads offered communication only between a few closely related cities, and how later the growth of the railroads brought into direct communication practically every village and metropolis throughout the land. Then came the time when the inadequacy of such complete but disconnected service struck the imagination of a man who saw the endless miles of track of countless railroads bound together by a supplemental system to which all railroads contributed and from which they profited, and by which, most of all, the public would enjoy a service of a scope which could otherwise only

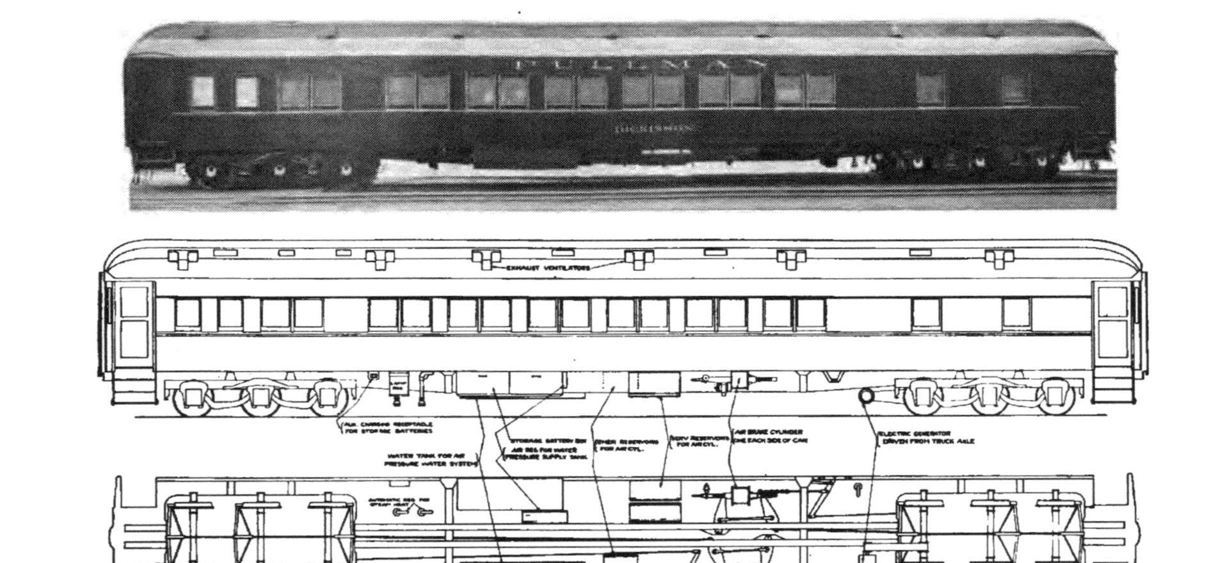

Pullman sleeping car, latest design, with outline drawing showing how the car is supplied with light, water and heat.

be attained by an actual combination of these railroads into a single company. But the vision of the founder of the Pullman Company did not stop at the idea of a unified system. He had not only seen the discomfort and inconvenience of countless changes from one train to another at railroad junctions and the midnight gatherings on the station platform; he had seen in tired eyes the fatigue of sleeplessness; he had seen in the preponderance of male passengers the lack of a protection sufficient to permit the free travel of unescorted women; he had realized, and his realization ranks high with the thoughts of the world's innovators, that travel was a hardship and that it could be made a pleasure.

With the realization constantly before him that the most perfect service could be given only by the most radically improved equipment and the widest extension of this company's activities, Mr. Pullman identified the early years of organization with a development of the passenger car to a degree of comfort, convenience, safety, and luxury that passed popular comprehension. Nothing was too good for the Pullman car; too much money could not be invested in it. Hand in hand with this development of the mechanical side of service he developed its extension throughout the country, by means of which it might be put into the hands of the greatest number of people for their greater convenience. Never has history more completely justified a business that from its character must be to a certain extent a monopoly. Never has competition more promptly yielded to unification.

It is natural to think of the Pullman Company as housed in some miraculous manner in the cars which it operates, as a company which expends its restless existence in untiring travel from state to state. But, as a matter of fact, the vast organization which makes possible the movement of the seventy-five hundred cars which comprise the present equipment holds an interest secondary only to the actual operation of the cars themselves.

There was a day when the run from Albany to Schenectady was the longest continuous railroad ride that a traveler might take. Today it is possible to travel in a Pullman car without change from Washington, D. C., to San Francisco, a distance of 3,625 miles, requiring one hundred and eighteen hours, or approximately five days.

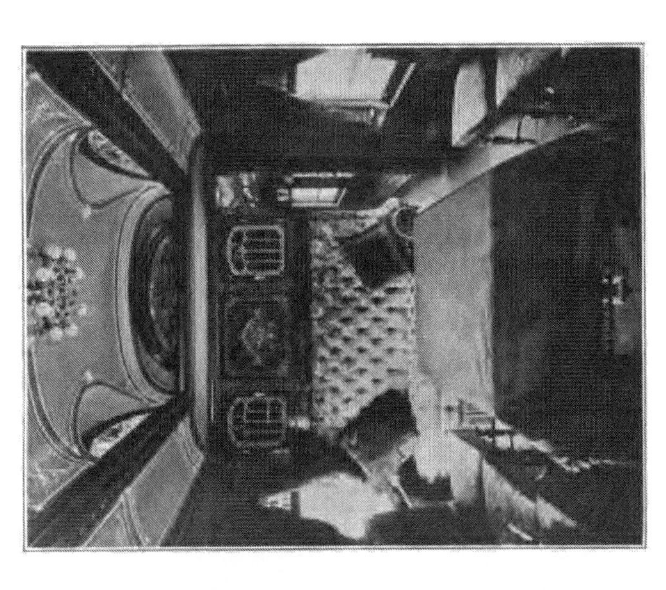

But distance is not alone characteristic of Pullman service; equal attention is given to shorter "hauls." From Greensboro to Raleigh, North Carolina, for instance, a distance of only eighty-one miles, Pullman sleeping cars are regularly operated. Here, as in many other instances, arrangements exist whereby the passengers may retire early in the evening while the car is at rest on a siding in the station, and arise at a reasonable hour in the morning. By such service hotel accommodations are practically afforded and it becomes possible for the travelers to have a whole day for pleasure or business at one place, spend a night in which a hundred or five hundred miles are traversed, and arrive without fatigue at another place the following morning.

The hotel desk corresponds to the ticket office of the Pullman Company. Imagine a hotel with 260,000 beds and 2,950 office desks, and a total registration of 26,000,000 people each year. This is what the Pullman Company does, however, and incidentally it does it often at a mile a minute and in every state in the Union. The 2,950 offices where Pullman berths, seats, drawing rooms or compartments may be purchased include Quebec,

Winnipeg, Manitoba, and Vancouver on the north; San Diego, El Paso, New Orleans, Key West, and Havana on the south; San Francisco on the west, and the seaboard towns of Maine on the east. Under normal conditions the southern limit is still further extended to fifty-six additional offices in the Republic of Mexico, as far south as Salina Cruz on the Gulf of Tehuantepec, and approximately two hundred miles from the boundary between Mexico and Guatemala, Central America.

The longest distance which it is possible to travel with a single Pullman ticket is from Portland, Maine, to San Francisco, by the way of Washington, D. C., New Orleans and Los Angeles. This cannot be done, however, in one sleeper, and changes must be made at New York and Washington. But a brief consideration of the perfect organization necessary to provide such continuous passage with berths reserved at each point of change by the mere purchase of a ticket at the starting point, grants to the Pullman Company a measure of credit due. In actual mileage the distance covered by this trip is 4,199.

As a rule the berths in sleeping cars and seats in

ROBERT T. LINCOLN
President of the Pullman Company from 1897 to 1911

parlor cars are on sale at the terminals of the different lines, but to provide facilities at intermediate points where the demand is sufficient to justify it, a limited number of sections are assigned for sale at such stations and tickets may be purchased from them on application. At stations of less importance and where the demand is not sufficient to assign any definite space, an arrangement exists whereby the vacant accommodations are telegraphed by ticket agents or conductors from point to point in order to accommodate passengers taking the trains at such stations. It is also possible and a very common practice to purchase a single sleeping car ticket between stations a great distance apart—for instance, between Boston, New York, Philadelphia, and Washington, to Los Angeles, San Francisco, Portland, and Seattle, via any of the ordinary routes of travel, by sufficient notice to the ticket agent to enable his reserving the accommodations, and it is also possible to purchase under similar conditions a sleeping car ticket in Havana, Cuba, for a berth, section, or drawing room from Key West, Florida, to Seattle, Washington, a distance of 3,923 miles, taking one hundred and thirty-three hours;

not, however, without change, but in connecting cars, giving continuous sleeping car service over various routes.

During the year 1916, 16,398,450 tickets of various forms were printed in Chicago and distributed to the various ticket offices, and in addition, 8,150,000 cash-fare tickets or checks were issued by conductors to travelers purchasing on the train.

In addition to offices where tickets may be purchased, arrangements exist in many thousands of smaller points whereby the public may secure sleeping-car accommodations by application to the station agent or other representative of the railroad company, who will arrange by telephone, telegraph, or letter the desired space to be called for, with a reasonable time at a designated point.

In order to extend to the public every courtesy consistent with lawful requirements and good business principles, the Pullman Company endeavors to provide prompt and careful attention to all requests for refund of fares where service paid for is not furnished, whether through the acts of its agents or employees or the passenger, or due to interruption of traffic.

Applications of this nature are usually made to the company's general offices in Chicago, but when this is not convenient, a report made to the company's representative in any of the important cities throughout the country is forwarded to the central offices and receives the most careful consideration.

It would seem of interest in this connection to state that during the year 1916, 53,743 applications, amounting to $152,446.00, were received for refund of fares, an average of one hundred and seventy-nine for each working day. Of the total number received 48,025 were considered favorably and paid, indicating the liberal policy of the company in such matters. Regardless of the amount involved, great or small, it is necessary that each case be considered on its individual merits, and the result determined with due regard to fairness to the passenger and the company, and not conflicting with legal necessities.

Probably seventy-five per cent of these requests for refunds are occasioned by passengers changing their plans or missing their train. Most frequent is the reason given that the wife has packed the tickets in the trunk, that the cab or taxi broke down, or that the last act of the theater caused unrealized delay. Often the tickets are lost, and not infrequently they are turned in by others for refund.

But one of the most convenient features of the Pullman service is the ease with which the traveler may reserve in advance accommodations on the train which he intends to take. In the ordinary railway coach it is a rule of "first come, first served" and the late arrival is often obliged to take a seat with a stranger. By the Pullman system, however, a call over the telephone or a stop at the local ticket office is all that is necessary to make as definite reservation of space as for a theater, and the traveler is wroth indeed when in rare instances a slip occurs and he finds his seat or berth has not been held for him and has been sold to another.

Naturally so general a convenience has led to rank abuses from which the passengers invariably suffer. Chief among them is the practice of hotel clerks and porters, especially in large cities and at summer and winter resorts, to reserve far in advance all the desirable Pullman accommodations on popular trains in the names of supposititious travelers whom they claim to represent, and later sell these

Bedroom and observation section of a costly private car. This car represents the apotheosis of railroad travel the case justifies the action. At a popular watering place an enterprising hotel employee figured out that on the day following Easter a large number of guests would leave on a certain popular train. Accordingly, like the theater "scalper," he purchased outright a large block of tickets on this train, in fact, every lower on the two Pullman sleepers. Fortunately the local agent of the company sensed that there was something "rotten in the state of Denmark" and made provision for two additional sleepers beyond the usual two which travel warranted. Being able to secure satisfactory accommodations direct from the agent the passengers failed to patronize the hotel porter's be-tipped and premiumed wares, and he, "stuck with the goods," tried a few days later to throw them back for refund on the Pullman Company. Their refusal cost him an even hundred dollars and broke up a peculiarly bad condition in that particular locality.

Many, indeed, are the difficulties attending the operation of a system of such magnitude, and it is only by a consideration of these difficulties that the true wonder of a service so nearly perfect can be appreciated.

OPERATING THE PULLMAN CAR

The operation of a system of such magnitude as the Pullman Company necessitates an operating organization letter perfect in its detail. Such an organization cannot be built to order; it must be a development, the result of years of wearying experience and costly experiment. In the introduction to the official book of instruction provided to car employees of the company, occurs, above the signature of the general superintendent, this sentence: "The most important feature to be observed at all times is to satisfy and please passengers." It is an apparently simple commission, a natural expression of desire, but a brief investigation of the requirements necessary "to satisfy and please" twenty-six million passengers, traveling rapidly from place to place, from north to south and from coast to coast, regardless of climate or locality, discloses a service and machinery for the carrying out of that service complete beyond the realization of the most discerning traveler.

To comprehend more clearly the details of this nation-wide service it must be considered in its two aspects — the material equipment which the operation of the cars requires, and the personal service afforded by the employees of the company. To give this service 7,500 cars of the Pullman Company are operated over one hundred and thirty-seven railroads, or a total of 223,489 miles of track, reaching practically every point in the country from which or to which a person might desire to travel. To operate these cars an army of over ten thousand car employees are required, while seven thousand more are employed to keep the cars in repair, and maintain them in a clean and sanitary condition.

The Pullman Company maintains, in addition to the great plant at Pullman, six repair shops situated at various convenient points throughout the country where cars are repaired and maintained in good condition. In 1916, a total of 5,115 cars were repaired at these various shops at a cost of over five million dollars. Only by such rigid maintenance can the cars be kept in the almost invariably excellent condition in which they are found by the public.

Years ago the wearied traveler wrapped his great coat about him for his midnight journey. Later a few "sleeping" cars of primitive construction provided sheets and blankets which were stored in a cupboard in the end of the car. As these were

Modern Pullman steel sleeping car, ready to be made up for the night

Modern Pullman steel sleeping car during the day

washed only at irregular intervals, it was a lucky passenger who found clean linen for his bed, and if he did not make up the bed himself, it was the brakeman who provided this domestic service. Naturally no one thought of undressing for the night, and when the Pullman car was first introduced it was necessary to print on the back of the tickets and in the employees' rules book the warning that passengers must not retire with their boots on.

Today the Pullman Company to provide clean linen nightly for each passenger, keeps on hand 1,858,178 sheets, which are valued at $980,553.00, and 1,403,354 pillow slips worth $186,475.00. In the twelve months ending April 27, 1916, over two hundred thousand sheets, valued at over one hundred thousand dollars, and nearly two hundred thousand pillow cases, valued at over twenty thousand dollars, were condemned. And during the same period 108,492,359 pieces of linen, including both sheets and pillow cases were washed and ironed. In the matter of condemnation, it is interesting to learn that the slightest tear or stain is considered sufficient cause. These figures are staggering in their immensity, but even more amazing is the system by which these articles are provided, changed, washed, returned in traveling hotels, at times hundreds of miles removed from the nearest supply station.

In the oldtime washroom a roller towel gave satisfaction to travelers less particular than those of the present day. But now how things have changed. Two million seven hundred thousand towels are needed to supply an ever increasing demand. Three hundred and twenty-five thousand dollars was their cost and each year seventy million towels is the laundry order. When Brown has shaved in the men's washroom in good American style, he will probably wipe his razor on a towel. It is not his custom at home, but the traveler seems to have scant respect for property. That one little cut will destroy the towel for future service. Pullman towels rarely have a chance to wear out. Over a hundred thousand a year are condemned chiefly because of such usage, and, sad to relate, each year over half a million are "lost." A Pullman towel is a handy wrapping for a pair of shoes, but the annual lost charge amounts to nearly seventy thousand dollars. It is a charge that must be accepted by the company. It will not do to question a passenger's integrity.

All told, the investment by the Pullman Company in car linen amounts to $1,856,708.00, representing 6,597,714 separate pieces. And this is only for sleeping and parlor cars and a relatively small number of buffet and private cars, for the company no longer operates the diners. To provide new linen to replace the lost and condemned costs an annual sum of over four hundred thousand dollars.

But the quantities and the cost of other articles which the company provides are even more impressive. These, for the most part, are expressions of Pullman service over and above the service itself, but it is unquestionably true that by such "over and above" service is the whole service most truly judged. Who would think, for instance, that in one year 5,819,656 women's hats were protected against dust by paper bags provided by the porters. And yet these paper bags represented a total cost of $14,549.00. Smokers in the same period consumed two million boxes of matches, and over forty-two million drinking cups costing nearly eighty thousand dollars gave the modern touch of sanitation to the water coolers. Soap would naturally be considered an essential part of the service, but a soap bill for one year of sixty thousand dollars is a large order for cleanliness. So, too, is the sum of $20,000 for hair brushes and a third of that amount for combs.

Back in the dark ages of blissful ignorance of germs, railroad coaches were hallowed breeding places for sickness. But times have changed, and today it is a pretty safe remark to make that the Pullman car is more healthful than almost any place where people frequently congregate. It does not take many gray hairs to remember the days of sleeping cars furnished with heavy carpets tacked to wooden floors, of stuffy hangings, and plush upholstery, of fancy woodwork rife with cracks and crannies, and of washrooms and toilets that no amount of cleaning could ever maintain entirely innocuous.

It is difficult to enumerate the countless little details that are constantly incorporated into Pullman car construction. The berth light has been frequently changed to embody some new idea to improve its convenience and efficiency. The coat hanger, and the mirror in the upper berth are minor details, but their convenience is attested by their constant use by passengers. In the washrooms the design of the wash basins has been frequently altered to afford a more convenient resting place for the toilet articles unpacked from the traveler's bag. Even the location of a coat hook receives a consideration that would perhaps seem exaggerated to the casual outsider. Double curtains are now provided on the newer cars, one set for the lower and another set for the upper berth.

Once a month a Committee on Standards, composed of the higher officials of the company, meets at the big plant at Pullman. On a track near the main entrance, stands a car in which every practical suggestion has been incorporated for the inspection of the committee. Some of these suggestions are quickly eliminated by their experienced verdict; others, possessing apparent worthiness, are passed and are later incorporated in the construction of the next cars manufactured, when the public will become the final judge. Many of these improvements are of a technical character, and primarily affect the construction of the cars; others are of a more directly personal nature and contribute more to the comfort and convenience of the traveler. All

tresses and pillows are hung in the open air for the action of that greatest of all purifiers, the sun. Blankets are given a similar treatment. Water coolers are cleaned and sterilized with steam. In fact, nothing that could harbor a speck of dust is neglected.

The slight, acrid odor sometimes noticeable in a Pullman car at the beginning of a run is caused by the disinfectants which are liberally employed. A jug of disinfectant solution is a part of the equipment of every car and this is used for all car washing and particularly on the floors and in the toilet and washrooms.

To protect still further the health of the passengers, the cars are regularly fumigated with a gas which kills all disease-producing bacteria. Whenever a car has carried a sick person it is fumigated as soon as it is vacated, in addition to the regular monthly, weekly, or other schedule of fumigation for various lines and terminals. In order that the district offices may be promptly informed as to the necessity of this extra fumigation, the conductor is required to note on his inspection report the fact that a sick passenger has been carried, and the car is immediately taken out of service and thoroughly cleaned and fumigated. Moreover, if space occupied by a sick passenger is vacated en route, it must not be resold until the car has reached its terminal and has been fumigated.

To provide the necessary facilities for car cleaning, the company maintains a cleaning force in two hundred and twenty-five principal yards, and, in addition, at one hundred and fifty-eight outlying points. These yards require the service of over four thousand cleaners.

Stationed throughout the United States, in nearly every city of prominence, are six superintendents, thirty-nine district superintendents and thirty agents. These men each week make personal inspection of cars in operation with the sole purpose of keeping the service up to the highest standard. In addition, a corps of electrical and mechanical inspectors constantly inspect and test the cars and their devices, at various places, and another corps of local inspectors carefully examine every departing and every incoming train with particular attention to the appearance and deportment of the car employees and the apparatus for heating, lighting and water.

OPERATING THE PULLMAN CAR

The Pullman Company is today the greatest single employer of colored labor in the world. Trained as a race by years of personal service in various capacities, and by nature adapted faithfully to perform their duties under circumstances which necessitate unfailing good nature, solicitude, and faithfulness, the Pullman porters occupy a unique place in the great fields of employment. There are porters who for over forty years have been employed by the company, and of all the porters employed, an army of nearly eight thousand, twenty-five per cent have been for over ten years in continuous service. The reputation of any company depends in a large measure on the character of its employees, and particularly in those concerns which render a personal service to the general public is it necessary that the standards of the employees be exceptionally high. Such standards of personal service cannot be quickly developed; they can be achieved only through years of experience and the close personal study of the wide range of requirements of those who are to be served.

To inspire in the car employees, conductors as well as porters, the ambition to satisfy and please the passenger, rewards of extra pay are made for unblemished records of courtesy; pensions are provided for the years that follow their retirement from active service; provision is made for sick relief, and at regular intervals increases in pay are awarded with respect to the number of years of continuous and satisfactory employment.

One characteristic of the Pullman business that is peculiarly significant is the average length of service of the employees. In a general way it may truly be said that from the car porter to the highest official every man who enters the business enters it as a life work. In most lines of business there is a variety of concerns operating along similar lines, and it is a natural step for a man to pass up from one company to another. But the unique position held by the Pullman Company has eliminated such a situation, and a man entering its employ looks forward to a personal development in this one concern.

During the half-century which has seen the sure and perfect development of this vast and complicated organization it is but natural to expect among the names of those who have guided its destiny many that must rank high in the business history of the

JOHN S. RUNNELLS
President of the Pullman Company

country. A glance at the list of past and present Directors of the company confirms the expectation. Here are the names of men who have found high places in a variety of business activities not only in Chicago but in other great cities. The list includes:

George M. Pullman
John Crerar
Norman Williams
Robert Harris
Thomas A. Scott
Amos T. Hall
C. G. Hammond
J. P. Morgan
Marshall Field
J. W. Doane
H. C. Hulbert
O. S. A. Sprague
Henry R. Reed
Norman B. Ream
William K. Vanderbilt
John S. Runnells
Frederick W. Vanderbilt
W. Seward Webb
Robert T. Lincoln
Frank O. Lowden
John J. Mitchell
Chauncey Keep
George F. Baker
John A. Spoor

In this same period but three men have occupied the office of president: George M. Pullman, the founder of the company, who held office from 1867, the year of incorporation, until his death in 1897, and Robert T. Lincoln until 1911, when John S. Runnells, the present president, was elected.

Pullman service has revolutionized the method of travel. Night has been abolished, the sense of distance has been annihilated; fatigue has been reduced to a minimum. In the oldest districts of the east, along the valleys of western rivers, on the widespread plains, among the remote peaks of the Rockies, in the deserts of the great southwest, the Pullman car, served by the same trained employees, furnishes the same comforts, and gives the same nights' repose. Improved each year in its mechanical construction, amplified in its service, better served by its attendants, it has set a high standard to the world in the development of railway travel, and in the fifty years of its development it has contributed more to the safety, comfort, convenience, and luxury of travelers than any other similar contribution that has been given to mankind.

INDEX

INDEX

INDEX

CPSIA information can be obtained
at www.ICGtesting.com
Printed in the USA
BVHW011812011221
623023BV00002B/74

9 781340 831363